March '81

Extraterrestrial
Civilizations

Isaac Asimov

Extraterrestrial Civilizations

Pan Books London and Sydney

First published in Great Britain 1980 by Robson Books Ltd
This edition published 1981 by Pan Books Ltd,
Cavaye Place, London SW10 9PG
© Isaac Asimov 1979
ISBN 0 330 26249 1
Printed and bound in Great Britain by
Richard Clay (The Chaucer Press) Ltd, Bungay, Suffolk

to the memory of Paul Nadan (1929–1978)
for whom I should have started the book sooner

Contents

1 The Earth

The question is: Are we alone?

Are human beings the only possessors of eyes that probe the depths of the Universe? The only builders of devices to extend the natural senses? The only owners of minds that strive to understand and interpret what is seen and sensed?

And the answer is, just possibly: We are not alone! There are other kinds that seek and wonder, and do so perhaps even more effectively than we.

Many astronomers believe this is so, and I believe this is so.

We don't know where those other minds are, but they are somewhere. We don't know what they do, but they do much. We don't know what they're like, but they are intelligent.

Will they find us if they are somewhere out there? Or have they found us already?

If they have not found us, can we find them? Better yet, *should* we find them? Is it safe?

These are the questions that must be asked once we agree that we are not alone, and astronomers are asking them.

The whole matter of the search for extraterrestrial intelligence has now become so common, in fact, that it has been abbreviated to save trouble in referring to it. Astronomers now refer to it as SETI, from the initials of the phrase 'the search for extraterrestrial intelligence'.

The first scientific discussion of SETI that offered a hope of carrying through the search successfully came only in 1959. It is natural to suppose, then, that the question of intelligence other than our own is of recent vintage. It would seem to be entirely a twentieth-century phenomenon arising out of the advance of astronomy in recent decades. It would seem to be the child of rocketry and of manned flight in outer space.

Perhaps you may feel that, prior to the last few decades,

human beings took it for granted that we *were* alone, and that the new view of other-intelligence is coming as a great shock to people and is forcing them willynilly to undergo an internal revolution of outlook.

Nothing could be farther from the truth!

It has been taken for granted by almost all people through almost all of history that we are *not* alone. The existence of other intelligences has been accepted as a matter of course.

Such beliefs have not arisen through the advance of science. Quite the contrary. What science has done has been to remove the supports from under the early casual assumptions as to the existence of other-intelligence. Science has created a new view of the world around us in which, by the old standards, humanity stands alone.

Let us start with the establishment of aloneness before we go on to a new view of a new kind of other-intelligence.

Spirits

To go back to the beginning, we will have to recognize that the phrase *extraterrestrial intelligence* is already sophisticated. It refers, after all, to intelligence found on worlds other than Earth and for it to have meaning there must be some recognition that worlds other than Earth exist.

To the vast majority of human beings, however, through almost all of history, there were no worlds other than Earth. Earth was *the* world, *the* home of living things. The sky, to early observers, was exactly what it appeared to be: a canopy overhanging *the* world, blue by day and punctuated by the round glare of the Sun; black by night and pinpricked with the brightness of the stars.

Under those conditions, the phrase *extraterrestrial intelligence* has no significance. Let us talk, instead of *nonhuman intelligence*.

As soon as we do that, we can see at once that human beings of the prescientific age always assumed that humanity was not alone; that the one world they thought of as filling the Universe contained a variety of nonhuman intelligences. Not only was human intelligence one of very many, but it was very likely to be the weakest and least advanced of all.

To the prescientific mind, after all, events in the world seemed whimsical and wilful. Nothing followed natural and inexorable 'law' because law was not recognized as part of the Universe. If something happened unpredictably, it was not because not enough was known to predict it, but because every part of the Universe was behaving with free will and doing things through some uncomprehended motivation – through even, perhaps, an incomprehensible motivation.

Free will is inevitably associated with intelligence. To do something wilful, after all, you have to understand the existence of alternatives and choose among them, and these are attributes of intelligence. It seemed to make sense, therefore, to consider intelligence a universal aspect of nature.

To the early Greeks (whose myths we know best), every aspect of nature had its spirits. Every mountain, every rock, every stream, every pool, every tree, had its nymph, marked not only by intelligence but even by a more or less human shape.

The ocean had its deity, as did the sky and the underworld; they were given human attributes such as childbirth and sleep, and various levels of abstraction such as art, beauty, and chance.

As time went on, Greek thinkers grew sophisticated enough to view all these spirits and deities as symbols, and to strive to withdraw them from human associations.

Thus, Zeus and his fellow gods were thought to live on Mount Olympus in northern Greece to begin with, but were later transferred to a vague 'Heaven' in the sky.* The same transfer took place in the case of the God of the Israelites, who originally lived on Mt Sinai or in the Ark of the Covenant, but who was eventually relocated to Heaven.

* Here was an example of another 'world' but one that was never visible or in any way sensed in the ordinary way.

In the same way, the world of the spirits of the dead could be thought of at first as sharing the one world with the living. Thus, in the *Odyssey*, Odysseus visits Hades in some vague spot in the far west, and it is somewhere in the west that the Elysian Fields, the Greek Paradise, may also have existed. The spirits of the dead were eventually transferred to a semi-mystical underground Hell.

Nevertheless, this process of sophisticated abstraction is a purely intellectual phenomenon intended to save the thinker the embarrassment of unsophisticated opinions. They rarely affected the common person.

Thus, whatever the Greek philosopher may have thought as to the cause of rain, the common uneducated farmer may have thought of rain (as Aristophanes jokingly says in one of his plays) as 'Zeus pissing through a sieve'.

In the contemporary United States, meteorology is a complex study, and the changes in weather are treated as natural phenomena that follow laws so complex, alas, that even yet we do not thoroughly understand them and can predict with only moderate accuracy. To many Americans, however, a drought, for instance, is the will of God, and they flock to the churches to pray for rain under the impression that the plans God has made are so trivial and unimportant that He will change them if asked to do so.

We are used to thinking of all the gods and demons of mythology as 'supernatural', but that is not really a proper use of the term. Any culture in its myth-making stage does not yet have the concept of natural law in the modern sense, so that nothing is really supernatural. The gods and demons are merely superhuman. They can do things that human beings cannot.

It is only modern science that introduced the concept of natural laws that cannot be broken under any circumstances – the various laws of conservation, the laws of thermodynamics, Maxwell's laws, quantum theory, relativity, the uncertainty principle, causal relationships.

To be superhuman is perfectly permissible, for cases of it are common. The horse is superhuman in speed; the elephant in strength; the tortoise in longevity; the camel in endurance; the dolphin in swimming. It is even conceivable that some non-human entity might be of superhuman intelligence.

To transcend the laws of nature, be 'supernatural' is, however, impermissible in the Universe as interpreted by science, in the 'Scientific Universe', which is the only one dealt with in this book.

It might easily be argued that human beings have no right to say that this or that is 'impermissible'; that something that is called supernatural receives its name by arbitrary definition out of knowledge that is finite and incomplete. Every scientist must admit that we do not know all the laws of nature that may exist, and that we do not thoroughly understand all the implications and limitations of the laws of nature that we think do exist. Beyond what little we know there may be much that seems 'supernatural' to our puny understanding, but that nevertheless *exists*.

Quite right, but consider this:

When we lead from ignorance, we can come to no conclusions. When we say, 'Anything can happen, and anything can be, because we know so little that we have no right to say *This is* or *This isn't*,' then all reasoning comes to a halt right there. We can eliminate nothing; we can assert nothing. All we can do is put words and thoughts together on the basis of intuition or faith or revelation and, unfortunately, no two people seem to share the same intuition or faith or revelation.

What we must do is set rules and place limits, however arbitrary these may seem to be. We then discover what we can say within these rules and limits.

The scientific view of the Universe is such as to admit only those phenomena that can, in one way or another, be observed in a fashion accessible to all, and to admit those generalizations (which we call laws of nature) that can be induced from those observations.

Thus, there are exactly four force fields that control all the interactions of subatomic particles and therefore, in the long run, all phenomena. These are, in order of discovery, the gravitational, the electromagnetic, the strong nuclear, and the weak nuclear interactions. No phenomenon that has been observed fails to be explained by one or another of these forces. No phenomenon is as yet so puzzling that scientists must conclude that some fifth force other than the four I've listed must exist.

It is perfectly possible to say that there is a fifth type of interaction that exists, but cannot be observed, or a sixth, or any number. If it cannot be observed, if it cannot make itself evident in any way, nothing is gained by talking about it – except, perhaps, for the amusement of inventing a fantasy.*

It is also perfectly possible to say that there is a fifth interaction (or a sixth or any number) that can indeed be observed, but only by certain people and only under certain unpredictable conditions.

That may conceivably be so, but it doesn't fall within the purview of science since under those conditions, *anything* can be said. I can say that the Rocky Mountains are made out of emeralds that have the property of looking like ordinary rock to everyone else but me. You can't disprove that statement but of what value is it? (Far from being of value, such statements are so annoying to people generally that anyone who insists on making them is liable to be treated as insane.)

Science deals only with phenomena that can be reproduced; observations that, under certain fixed conditions, can be made by anybody of normal intelligence; observations upon which reasonable men can agree.†

It may well be argued, in fact, that science is the only field of human intellectual endeavour on which reasonable men can

---

* I do not wish to denigrate the value of inventing fantasy. It is a noble art, requiring great skill. I know. I have been making my living out of it for years. It is one thing, however, to invent an amusing fantasy, and quite another to confuse it with reality.

† I won't bother trying to define a 'reasonable man'. I suspect that one convenient assumption we can make is that anyone bothering to read this book is a reasonable man.

very often agree, and in which reasonable men can sometimes change their minds as new evidence comes in. In politics, art, literature, music, philosophy, religion, economics, history – carry on the list as long as you wish – otherwise reasonable men can not only disagree, but invariably do, and sometimes with the utmost passion; and never change their minds, either, it would appear.

Of course, the scientific world view has not been handed down intact from time immemorial. It was discovered and worked out little by little. It is not complete now, and it may never be entirely complete. New refinements, modifications, additions may seem fantasy at first (quantum theory and relativity certainly did), but there are well-known ways of testing such things carefully; and if the theories pass, they are accepted. The testing method is not always simple and easy, and in the course of the testing disputations may arise and verification may be unnecessarily delayed.*

Acceptance will come in the end, though, for scientific thought is self-correcting as long as there is reasonable freedom of research and publication. (Without infinite money and infinite space, it is hard to be sure of *absolute* freedom, of course.)

All this is my justification for having this book deal with the supernormal whenever necessary, but never with the supernatural. In the discussion of nonhuman intelligence that will occupy us in this book, we will consider neither angels nor demons, neither God nor Devil, nor anything that is not accessible to observation and experiment and reason.

* Such disputation can be quite nasty and polemical at times, for scientists are quite human, and any given individual among then can be, at times, petty, mean, vindictive – or simply stupid.

Animals

In our search for nonhuman intelligence on Earth, then, having eliminated all the wonderful things the human imagination has constructed out of nothing, we must find what we can in the dull things we can sense and observe.

Of the natural objects on Earth, we can, in our search for intelligence, at once eliminate the inanimate, or nonliving ones.

This is by no means an indisputable decision, for it is not an impossible thought that consciousness and intelligence are inherent in all matter; that individual atoms, even, have a certain micro-quantity of such things.

That may be so, but since such consciousness or intelligence cannot (as yet, at least, and we have no choice but to go with the 'as yet') be in any way measured, or even observed, it falls outside the Universe as I intend to deal with it, and we can eliminate it.

Besides, if we are looking for nonhuman intelligence, it may be taken for granted that we are seeking for intelligence that, while present in something other than a human being, is nevertheless at least roughly comparable in quality to intelligence in a human being. That means it must be intelligence we can clearly recognize as such, and whatever intelligence there may be in a rock, it is not the kind of intelligence we can recognize.

Ah, but must all kinds of intelligence be the same, or even similar, or even recognizable? Might not a boulder be as intensely intelligent, as we are, or more intensely, but be so in a completely unrecognizable way?

If that is so, there is nothing to prevent us from saying that every individual object in the entire Universe is as intelligent as a human being, or more intelligent than one, but that in the case of every single one of those objects, the nature of the intelligence is so different from ours as to be unrecognizable.

If we can successfully maintain that, all argument stops right there and there is no room for further investigation. We *must* set limits, if we are to continue. In searching for nonhuman intelligence, we can reasonably limit ourselves to such intelligence that we can recognize as such (even if only dimly) from reproducible observations and by using our own intelligence as a standard.

It is possible that intelligence may be so different from ours that we don't recognize it at once, but do come to recognize it by degrees. However, in all the years of human association with inanimate objects, there has been no real reason to suppose any of them to have shown any sign of intelligence, however small, and it is as reasonable as anything can well be to eliminate them.*

If we pass on to animate objects, we might next raise the question of how to distinguish between inanimate and animate objects. The distinction is harder than we might think, but it is irrelevant. All those objects that offer the slightest chance of confusion as to their classification, whether living or nonliving, clearly do not represent reasonable claims to the possession of nonhuman intelligence.

And of those objects that are indisputably living, we can eliminate the entire plant world. There is no recognizable intelligence in the most magnificent redwood, the sweetest-smelling rose, the most ferocious Venus'-flytrap.†

When it comes to animals, however, matters are different. Animals move as we do and have recognizable needs and fears as we do. They eat, sleep, eliminate, reproduce, seek comfort, and avoid danger. Because of this, there is a tendency to read

* I make an exception of those inanimate objects, called computers, that have come to exist in the last quarter-century, and that, in some ways, give evidence of properties that can easily be mistaken for intelligence. These are, however, human products, and can fairly be considered as extensions of human intelligence, and not as nonhuman intelligence.

† There are books that have been written describing how plants seem aware of human speech and react with apparent intelligence to it. As far as biologists can tell, however, there is no scientific merit whatever to such views.

into their actions human motivation and human intelligence.

Thus, to the human imagination, ants and bees, which follow behaviour that is purely instinctive and with little or no scope for individual variation, or for behaviour change to meet unlooked-for eventualities, are viewed as being purposefully industrious.

The snake, which slithers through the grass because that is the only way its evolved shape and structure makes it possible for it to move, and which thus avoids notice and can strike before being seen, is imagined to be sly and subtle. (This characterization can be upheld on the authority of the Bible – see Genesis iii, 1.)

In similar fashion, the donkey is thought of as stupid, the lion and eagle as proud and regal, the peacock as vain, the fox as cunning, and so on.

It is almost inevitable that wholesale attribution of human motivations to animal actions will lead one to take it for granted that if one could but establish communication with particular animals one would find them of human intelligence.

This is not to say that particular human beings, if pinned to the wall, will admit believing this. Nevertheless, we can watch Disney cartoons featuring animals with human intelligence and remain comfortably unaware of the incongruence.

Of course, such cartoons are just an amusing game, and the willing suspension of disbelief is a well-known characteristic of human beings. Then, too, Aesop's fables and the medieval chronicles of Reynard the Fox are not really about talking animals, but are ways of expressing truths about social abuses without risking the displeasure of those in power – who may not be bright enough to recognize that they are being satirized.

Nevertheless, the enduring popularity of these animal stories, to which one can add Joel Chandler Harris's 'Uncle Remus' tales and Hugh Lofting's 'Dr Dolittle' stories, shows a certain readiness in the human being to suspend disbelief in that particular direction; more so, perhaps, than in others. There is a sneaking feeling, I suspect, that if animals aren't as intelligent as we are, they ought to be.

We cannot even seek refuge in the fact that talking-animal stories are essentially for children. The recent best-sellerdom of *Watership Down* by Richard Adams is an example of a talking-animal book for adults that I found profoundly moving.

And yet, side by side with this ancient and primordial feeling of cousinship with animals (even while we hunted them down or enslaved them) there is, in Western thought at least, the consciousness of an impassable gulf between human beings and other animals.

In the Biblical account of creation, the human being is created by God through an act different from that which created the rest of the animals. The human being is described as created in God's image and as being given dominion over the rest of creation.

This difference can be interpreted as meaning that the human being has a soul and that other animals do not; that there is a spark of divinity and immortality in human beings that is not present in other animals; that there is in human beings something that will survive death, while nothing of the sort is present in other animals.

All this falls outside the purview of science and can be disregarded. The influence of such religious views, however, makes it easier to believe that human beings alone are reasoning entities and that no other animal is. This, at least, is something that can be tested and observed by the usual methods of science.

Nevertheless, human beings have not been secure enough in the uniqueness of our species to be willing to let it stand the test of scientific investigation. There has even been a certain nervousness about the tendency of those biologists with a strong concept of order to classify living things into species, genera, orders, families, and so on.

By grouping animals according to greater and lesser resemblances, one develops a kind of tree of life with different species occupying different twigs of different branches. What starts out as an inescapable metaphor suggests only too clearly the possibility that the tree grew; that the branches developed.

In short, the mere classification of species leads inexorably to the suspicion that life evolved; that more intelligent species, for instance, developed from less intelligent ones; and that, in particular, human beings developed from primitive species that lacked the capacities we now consider peculiarly human.

Indeed, when Charles Darwin published his *On the Origin of Species* in 1859, there was an outburst of anger against it, even though Darwin carefully avoided discussing human evolution. (It was to be another decade before he dared publish *The Descent of Man.*)

To this day, many people find it difficult to accept the fact of evolution. They don't, apparently, find the suggestion offensive that there are human characteristics in animals such as mice (who can be more lovable than Mickey?), but they do find it offensive that we ourselves may be descended from subhuman ancestors.

Primates

In the classification of animals there is an order called Primates, which includes those popularly known as monkeys and apes. In their appearance the primates resemble the human being more than any other animals do, and from that appearance it is natural to deduce that they are more closely related to human beings than other animals are. In fact, the human being must be included as a primate, if any sense at all is to be made of animal classification.

Once evolution is accepted, one must come to the inevitable conclusion that the various primates, *including the human being,* have developed from some single ancestral stem and that all are to varying degrees cousins, so to speak.

The resemblance of other primates to human beings is both endearing and repulsive. The monkey house is always the most popular exhibit in a zoo, and people will watch anthropoid apes (which most closely resemble the human being) with fascination.

The English dramatist William Congreve wrote in 1695, however, 'I could never look long upon a monkey, without very mortifying reflections.' It is not hard to guess that those 'mortifying reflections' must have been to the effect that human beings might be described as large and somewhat more intelligent monkeys.

Those who oppose the idea of evolution are often particularly hard on apes, exaggerating their nonhuman characteristics in order to make less likely any notion of kinship between them and ourselves.

Anatomical distinctions were sought, some little bodily structure that might be present in human beings alone and not in other animals, and most particularly not in apes. None has ever been found.

In fact, the superficial resemblance between ourselves and other primates, and in particular between ourselves and the chimpanzee and gorilla, becomes all the deeper on closer examination. There is no internal structure present in the human being that is not also present in the chimpanzee and gorilla. All differences are in degree, never in kind.

But if anatomy fails to establish an absolute gulf between human beings and the most closely related nonhuman animals, perhaps behaviour can do so.

For instance, a chimpanzee cannot talk. Efforts to teach young chimpanzees to talk, however patient, skilful, and prolonged those efforts may be, have always failed. And without speech, the chimpanzee remains nothing but an animal. (The phrase *dumb animal* does not refer to the lack of intelligence of the animal, but to its muteness, its inability to speak.)

But might it be that we are confusing communication with speech?

Speech is, we may take for granted, the most effective and delicate form of communication of which we are aware, but is it the only one?

Human speech depends upon human ability to control rapid and delicate movements of throat, mouth, tongue, and lips, and all this seems to be under the control of a portion of the brain called Broca's convolution, named for the French surgeon Pierre Paul Broca (1824–1880). If Broca's convolution is damaged by a tumour or a blow, a human being suffers from aphasia and can neither speak nor understand speech. Yet such a human being retains intelligence and is able to make himself understood, by gesture for instance.

The section of the chimpanzee's brain equivalent to Broca's convolution is not large enough or complex enough to make speech in the human sense possible. But what about gesture? Chimpanzees use gestures to communicate in the wild; could that use be improved?

In June 1966, Beatrice and Allen Gardner of the University of Nevada chose a one-and-a-half-year-old female chimpanzee they named Washoe and decided to try to teach her a deaf-and-dumb language of gestures. The results amazed them and the world.

Washoe readily learned dozens of signs, using them appropriately to communicate desires and abstractions. She invented new modifications, which she also used appropriately. She tried to teach the language to other chimpanzees and she clearly enjoyed communicating.

Other chimpanzees have been similarly trained. Some have been taught to arrange and rearrange magnetized counters on a wall. In so doing, they showed themselves capable of taking grammar into account and were not fooled when their teachers deliberately created nonsense sentences.

Young gorillas have been similarly trained and have shown even greater aptitude than chimpanzees.

Nor is it a matter of conditioned reflexes. Every bit of evidence shows that chimpanzees and gorillas know what they are doing, in the same sense that human beings know what they are doing when they talk.

To be sure, the ape language is very simple compared to the language of human beings. The human being is enormously more intelligent than apes, but again the difference here is one of degree rather than kind.

Brains

To anyone considering the comparative intelligence of animals, it is clear that the key anatomical factor is the brain. Primates have larger brains in general than the large majority of non-primates, and the human brain is the largest primate brain by a good deal.

The brain of an adult chimpanzee weighs 380 grams ($13\frac{1}{2}$ ounces) and that of an adult gorilla weighs 540 grams (19 ounces or just under $1\frac{1}{4}$ pounds). In comparison, the brain of an adult male human being weighs on the average 1,450 grams ($3\frac{1}{4}$ pounds).

The human brain is not, however, the largest that has ever evolved. The largest elephants have brains as massive as 6,000 grams (about 13 pounds) and the largest whales have brains that reach a mark of 9,000 grams (nearly 19 pounds).

There is no question but that the elephant is among the more intelligent animals. In fact, the intelligence of the elephant is so apparent that human beings tend to exaggerate it. (There is a greater tendency to exaggerate the elephant's intelligence than the ape's, perhaps because the elephant is so different from us in appearance that it represents a lesser threat to our uniqueness.)

We do not have the opportunity to study whales as we do elephants, but we may readily believe that whales are among the more intelligent animals, too.

Yet, although elephants and whales are relatively intelligent, it is quite clear that they are far less intelligent than human beings, and may well be less intelligent than the chimpanzee and gorilla. How may this be squared with the superhuman size of their brains?

The brain is not merely an organ of intelligence; it is also the medium through which the physical aspects of the body are organized and controlled. If the physical size of the body is great, enough of the brain is occupied with the physical to allow little for the purely intellectual.

Thus, each pound of chimpanzee brain is in charge of 150 pounds of chimpanzee body, so that the brain–body ratio is 1:150. In the gorilla, the ratio may be as low as 1:500. In the human being, on the other hand, the ratio is about 1:50.

Compare this with the elephant, where the brain–body ratio is as little as 1:1,000 and the largest whales, with as little as 1:10,000. Now it is not so surprising that there is something special about human beings that the large-brained elephants and whales do not seem to duplicate.

Yet there are organisms in which the brain–body ratio is actually more favourable than in the human being. This is true for some of the smaller monkeys and for some of the hummingbirds. In some monkeys the ratio is as great as 1:17.5. Here, though, the absolute mass of the brain is too small to carry much of an intellectual load.

The human being strikes a happy medium. The human brain is large enough to allow for high intelligence; and the human body is small enough to allow the brain space for intellectual endeavour.

Yet even here the human being does not stand alone.

In considering the intelligence of whales, it is perhaps not fair to deal with the largest specimens. One might as well try to gauge the intelligence of primates by considering the largest member, the gorilla, and ignoring its smaller cousin, the human being.

What of the dolphins and porpoises, which are pygmy relatives of the gigantic whales? Some of these are no more massive than human beings and yet have brains that are larger than the human brain (with weights up to 1,700 grams, or $3\frac{3}{4}$ pounds) and more extensively convoluted.

It is not safe to say from this alone that the dolphin is more intelligent than the human being, because there is the question of the internal organization of the brain. The dolphin's brain may be organized for predominantly nonintellectual purposes.

The only way to tell is to study dolphin behaviour, and here we are sadly hampered. They seem to communicate by modulated sounds even more complicated than those of human languages, yet we can make no progress in understanding dolphin communication. They seem to show signs of intelligent behaviour, even kindly and humane behaviour, yet on the other hand their environment is so different from ours that it is difficult for us to get inside their skin and grasp their thoughts and motivations.

The question of the exact level of dolphin intelligence remains, at least for now, moot.

Fire

In the light of the previous sections of this chapter, the question as to whether nonhuman intelligence exists on Earth must be answered: Yes.

It would seem that my contention early in the chapter that science has made us alone has not been demonstrated. There are a number of animals with surprisingly high intelligence, and these include not only apes, elephants, and dolphins. Crows are surprisingly intelligent when compared with other birds, and octopi show a level of intelligence far surpassing that of other invertebrates.

And yet absolute differences *do* exist; unbridgeable gulfs *are* there. The clue lies not so much in the mere presence of intelligence but in what is done through the use of that intelligence.

Human beings have been defined as tool-making animals and, to be sure, even the small-brained hominids who were our precursors were already making use of shaped pebbles a couple of million years ago. This is not surprising, since even the small-brained hominids had brains that were rather better than those of the apes of today.

However, other animals, even some who are quite unintelligent, make use of stones and twigs in ways that can only be considered as tool using.

It is not, then, tool making in itself that establishes a clear gulf between the human being and other intelligent animals.

And yet there may be some one kind of tool that marks the clear boundary line separating the most intelligent species from all others.

We have not far to seek. The key lies in the taming and use of fire. There is definite evidence of fire's having been used in caves in China in which an earlier hominid species, *Homo erectus*, dwelt at least half a million years ago. The discovery has never been forgotten.

No human society existing anywhere on Earth now lacks the knowledge of how to ignite and use a fire. No nonhuman species whatever has ever made the slightest advance in the direction of the use of fire, as far as we can tell.

Suppose we define 'human intelligence' as: a level of intelligence high enough to allow the development of methods for igniting and using fire.

In that case, to the question of whether the equivalent of human intelligence exists on Earth in nonhuman species, the answer must be: No! – The human being stands alone.

This might seem unfair; and the result of an arbitrary, self-serving definition. Let's see if it is by comparing the dolphin and the human being.

The dolphin spends his life in water and the human being spends his life in air. Water is a viscous medium, much more viscous than air. It takes much more effort to force one's way through water at a given speed than it does through air. (Anyone who has tried to run when partly immersed in water knows this is so.)

In order to attain speed in water, the dolphin has evolved a streamlined form to reduce water resistance. Moving through air, however, the human being does not require streamlining. The human being can develop a very irregular form and still be capable of fast motion.

For that reason, the human being can develop complicated appendages, while the dolphin cannot. The dolphin's streamlining allows it two stubby paddles and a fluke as its only manoeuvrable appendages, and these are useful only for propulsion and guiding.

To put it most briefly, human beings, because they live in air, can develop hands with which they can manipulate their environment. Dolphins, because they live in water, cannot develop hands.

Then again, the fire that early humans learned to handle is the radiation of heat and light that results from a rapid energy-yielding chemical reaction. The most common energy-yielding large-scale chemical reactions that are useful in this connection are those resulting from the combination of substances containing carbon atoms, hydrogen atoms, or both ('fuel') with the oxygen in the air. The process is called combustion. Fire cannot exist under water since free oxygen is not present and combustion cannot take place.

Therefore, even if dolphins had the intelligence to conceptualize fire, and to work out, mentally, the steps needed to tame and use it, they would be unable to put any of it into practice.

We see now, however, that the human use of fire could be considered as no more than the accidental by-product of the fact that the human being lives in air, and is not in itself necessarily a true measure of intelligence.

The dolphins, after all, even though they are unable to manipulate the environment and unable to build and use a fire, may have in their own way developed a subtle philosophy of life. They may have worked out, more usefully than we have, a rationalization of living. They may interchange more joy and good will with their feelings and understand more. The fact that we cannot grasp their philosophy and their moods of thought is no evidence of their low intelligence, but is perhaps evidence of our own.

Well, *perhaps*!

The fact is, though, that we don't have any evidence of the dolphin's philosophy of life. The lack of that evidence may be entirely our fault, but there's nothing we can do about it. Without evidence, there is no way of reasoning usefully. We can look for the evidence and some day, perhaps, find it, but until then, we can't reasonably assign human intelligence to the dolphin.

Besides, even if our definition of human intelligence on the basis of fire is unfair and self-serving on some abstract scale, it will prove useful and reasonable for the purposes of this book. Fire sets us on a road that ends with a search for extraterrestrial intelligence; without fire we would never have made it.

The extraterrestrial intelligences we are looking for, then, must have developed the use of fire (or, to be fair, its equivalent) at some time in their history, or, as we are about to see, they could not have developed those attributes that would make it possible for them to be detected.

Civilization

Throughout the history of life, species of living creatures have made use of chemical energy by the slow combination of certain chemicals with oxygen within their cells. The process is analogous to combustion, but is slower and much more delicately controlled. Sometimes use is made of energy available in the bodies of stronger species as when a remora hitches a ride on a shark, or a human being hitches an ox to a plough.

Inanimate sources of energy are sometimes used when species allow themselves to be carried or moved by wind or by water currents. In those cases, though, the inanimate source of energy must be accepted at the place and time that it happens to be and in the amount that happens to exist.

The human use of fire involved an inanimate source of energy that was portable and could be used wherever desired. It could be ignited or extinguished at will and could be used when desired. It could be kept small or fed till it was large, and could be used in the quantities desired.

The use of fire made it possible for human beings, evolutionarily equipped for mild weather only, to penetrate the temperate zones. It made it possible for them to survive cold nights and long winters, to achieve security against fire-avoiding predators, and to roast meat and grain, thus broadening their diet and limiting the danger of bacterial and parasitic infestation.

Human beings multiplied in number and that meant there were more brains to plan future advances. With fire, life was not quite so hand-to-mouth, and there was more time to put those brains to work on something other than immediate emergencies.

In short, the use of fire put into motion an accelerating series of technological advances.

About 10,000 years ago, in the Middle East, a series of crucial advances were made. These included the development of agriculture, herding, cities, pottery, metallurgy, and writing. The final step, that of writing, took place in the Middle East about 5,000 years ago.

This complex of changes stretching over a period of 5,000 years introduced what we call civilization, the name we give to a settled life, to a complex society in which human beings are specialized for various tasks.

To be sure, other animals can build complex societies and can be composed of different types of individuals specialized for different tasks. This is most marked in such social insects as bees, ants, and termites, where individuals are in some cases physiologically specialized to the point where they cannot eat, but must be fed by others. Some species of ants practice agriculture and grow small mushroom gardens, while others herd aphids; still others war on and enslave smaller species of ants. And, of course, the beehive and the ant or termite colony have many points of analogy with the human city.

The most complex nonhuman societies, those of the insects, are, however, the result of instinctive behaviour, the guidelines of which are built into the genes and nervous systems of the individuals at birth. Nor does any nonhuman society make use of fire. With insignificant exceptions, insect societies are run by the energy produced by the insect body.

It is fair, then, to consider human societies as basically different from other societies and to attribute what we call civilization to human societies only.

A third group of changes began about 200 years ago with the development of a practical steam engine, leading on to an Industrial Revolution, which is still in progress. And about twenty years ago we began to dispose of types of energy that could leak out into space in noticeable quantities. *We* became detectable.

In short, we are not looking merely for extraterrestrial life. We are not even looking merely for extraterrestrial intelligence. We are looking for extraterrestrial civilization that disposes of enough energy of a sufficiently sophisticated kind to be detectable over interstellar distances. After all, if the level of life/intelligence/civilization on some world is such that it is indetectable, we are not going to detect it.

And now, you see, it is fair to say that on Earth there is exactly one civilization of the kind we are looking for; just one, our own. As far as we know, there has never been any other civilization of this kind on Earth, and it was only a few years ago that our own civilization became the kind I'm referring to – a detectable civilization.

Of course, now that I've demonstrated that, in our role of civilization-makers, we are alone on Earth – that is no great tragedy after all. Earth is no longer the only world in the consciousness of human beings. We need only look for civilizations elsewhere, on other worlds, and it may be discovered that we are not alone after all.

2 The Moon

Phases

If we imagine ourselves looking about at our surroundings with no knowledge concerning them at all, we might be forgiven for thinking the Earth was the only world there was. What, then, made people think there were other worlds?

It was the Moon. Consider:

The predominant characteristic of the objects in the sky is their glow. The stars are little bits of sparkling light. The planets are somewhat brighter bits of sparkling light. The Sun is a round circle of blazing light. There is an occasional meteor that produces a brief line of light. There is an even more occasional comet that is an irregular hazy patch of light.

It is the light that makes the heavenly objects seem altogether different from the Earth, which in itself is dark and gives off no light.

To be sure, light can be produced on the Earth in the form of fire, but that is altogether different from the heavenly light. Earthly fires have to be fed constantly with fuel or they would flicker and go out, but the heavenly light continues for ever without change.

In fact, the Greek philosopher Aristotle (384–322 BC) maintained that all the heavenly objects were composed of a substance called aether, separate and distinct from the elements that made up the Earth. The word *aether* is from the Greek word meaning *to blaze*. The heavenly objects blazed and the Earth did not, and as long as that was thought to be true there was only one world; one solid, dark object on which life could exist, and many blazing objects on which life could not exist.

Except that there is the Moon. The Moon is the one heavenly object that changes shape in a regular way and in a fashion that is clearly visible to the unaided eye. These different shapes of the Moon (its 'phases') are ideally suited to attract attention and, except for the succession of day and night, were probably the first astronomical changes to catch the attention of primitive human beings.

The Moon goes through its complete cycle of phases in a little over twenty-nine days, which is a particularly convenient length of time. To the prehistoric farmer and hunter, the cycle of seasons (the year) was very important, but it was difficult to note that, on the average, the seasons repeated themselves every 365 or 366 days. The number was too large to be kept track of easily. To count 29 or 30 days from each new Moon to the next, and then to count 12 or 13 new Moons to each year, was much simpler and much more practical. The making of a calendar that would serve to keep track of the seasons of the year in terms of the phases of the Moon was a natural result of very early astronomical observations.

Alexander Marshak, in his book *The Roots of Civilization*, published in 1972, argues persuasively that, long before the beginning of recorded history, early human beings were marking stones in a code designed to keep track of the new Moons. Gerald Hawkins, in *Stonehenge Decoded*, argues just as persuasively that Stonehenge was a prehistoric observatory also designed to keep track of the new Moon, and to predict the lunar eclipses that occasionally came at the time of the full Moon. (A lunar eclipse was a frightening 'death' of the Moon upon which human beings depended for keeping track of the seasons. To be able to predict its occurrence reduced the fear.)

It was very likely the overriding practical necessity of working out a calendar based on the phases of the Moon that forced human beings into astronomy, and from that to a careful observation of natural phenomena generally, and from that to the eventual growth of science.

The fact that the phase changes were so useful could not help, it seems to me, but reinforce the notion of the existence of a benevolent deity who, out of his love of humanity, had arranged the skies into a calendar that would guide mankind into the proper ways of ensuring a secure food supply.

Each new Moon was celebrated as a religious festival in many early cultures, and the care of the calendar was usually placed in priestly hands. The very word *calendar* is from the Latin word meaning *to proclaim*, since each month only began when the coming of the new Moon was officially proclaimed by the priests. We could conclude, then, that a considerable portion of the religious development of mankind, of the belief in God as a benevolent parent rather than a capricious tyrant, can be traced back to the changing face of the Moon.

In addition, the fact that close study of the Moon was so important in controlling the daily lives of human beings could not help but give rise to the notion that the other heavenly objects might be important in this respect, also. The face of the Moon may in this way have contributed to the growth of astrology and, thereby, of other forms of mysticism.

But in addition to all this (and it would scarcely seem that if the Moon has given rise to science, religion, and mysticism, more should be required of it) the Moon gave rise to the concept of the plurality of worlds – the notion that the Earth was only one world of many.

When human beings first stared at the Moon from night to night in order to follow its phases, it was natural to suppose that the Moon literally changed shape. It was born as a thin crescent, waxed to a full circle of light, then waned to a crescent and died. Each new Moon was literally a *new* Moon, a fresh creation.

Quite early on, however, it became apparent that the horns of the lunar crescent *always* faced away from the Sun. That alone was sufficient to indicate some connection between the Sun and Moon's phases. Once the notion of that connection arose, further observation would show that the phases were connected with the relative positions of the Sun and Moon. The Moon was full when it and the Sun were at precisely opposite parts of the sky. The Moon was in the half-phase when it and the Sun were separated by ninety degrees. The

Moon was in crescent shape when it was close to the Sun, and so on.

It became apparent that if the Moon were a sphere that was as dark as the Earth, and if the Moon shone only by the light that fell upon it from the Sun and was reflected by it, then it would go through precisely the cycle of phases that were actually observed. The idea arose and grew to be more and more accepted that the Moon, at least, was a dark body like Earth and was not composed of blazing 'aether'.

Another world

If the Moon were like the Earth in being dark, might it not be like the Earth in other ways? Might it not be a second world?

As early as the fifth century BC, the Greek philosopher Anaxagoras (500–428 BC) expressed his opinion that the Moon was an Earthlike world.

To imagine the Universe as consisting of one world plus bits of light is intellectually acceptable. To imagine it to consist of two worlds plus bits of light is difficult. If one of the objects in the sky is a world, why not some or all the rest? Gradually, the notion of the plurality of worlds spread. Increasing numbers of people began to think of the Universe as containing many worlds.

But not *empty* worlds. That thought apparently filled people with revulsion – if it occurred to them at all.

The one world we knew – Earth – is full of life, and it is only natural to think that life is as inevitable a characteristic of worlds generally as solidity is. Again, if one thinks of the Earth as having been created by some deity or deities, then it is logical to suppose the other worlds to have been so created as well. It would then seem nonsensical to suppose that any world would be created and left empty. What motivation could there be in creating empty worlds? What a waste it would be!

Thus, when Anaxagoras stated his belief that the Moon was an Earthlike world, he also suggested that it might be inhabited. So did other ancient thinkers, as for instance the Greek biographer Plutarch (AD 46–120).

Then again, if a world is inhabited, it seems natural to suppose it to be inhabited by intelligent creatures – usually pictured as very much like human beings. To suppose a world to be inhabited only by unreasoning plants and animals would, again, seem to represent an intolerable waste.

Oddly enough, there was talk of life on the Moon even before the Moon was recognized as a world. This arose out of the fact that the Moon is again unique among heavenly bodies in not being evenly shining. There are darker smudges against the bright light of the Moon, smudges that are most clearly and dramatically visible at the time of the full Moon.

It was tempting for the average unsophisticated observer of the Moon to try to make a picture out of the smudges upon its face. (In fact, even the sophisticated and knowledgeable present-day observer may be tempted to do so.)

Given the natural anthropocentricity of human beings, it was almost inevitable that those smudges were pictured as representing a human being, and the notion of the 'man in the Moon' arose.

Undoubtedly the original notion was prehistoric. In medieval times, however, attempts were often made to clothe age-old notions with a cloak of Biblical respectability. Therefore, the man in the Moon was thought to represent the man mentioned in Numbers xv, 32–36: 'And while the children of Israel were in the wilderness they found a man that gathered sticks upon the sabbath day ... And the Lord said unto Moses, the man shall be surely put to death ... And all the congregation brought him without the camp, and stoned him with stones, and he died ...'

There is no mention of the Moon in the Biblical story, but it was easy to add the tale that when the man protested that he did not want to keep 'Sunday' on Earth (although to the Israelites, sabbath fell on the day we call Saturday), the judges said, 'Then you shall keep an eternal Monday [Moon-day] in heaven.'

The man in the Moon was pictured in medieval times as bearing a thornbush, representing the sticks he had gathered; and a lantern, for he was supposed to have been gathering them at night when he hoped no one would see; and, for some reason, a dog. The man in the Moon, with these appurtenances, is part of the play within a play presented by Bottom and the other rustics in William Shakespeare's *A Midsummer Night's Dream.*

Of course, the man in the Moon was visualized as filling his entire world, since the smudges seemed smeared over the entire face of the Moon, and since the Moon appears to be a small object.

It was the Greek astronomer Hipparchus (190–120 BC) who first managed to work out the size of the Moon relative to the Earth by valid mathematical methods and who got essentially the right answer. The Moon is an object about one-quarter the diameter of the Earth. It was no man-in-the-Moon-sized object. It was a world not only in the dark nature of the material making it up, but in its size.

What's more, Hipparchus had worked out the distance to the Moon. It is sixty times as far from the surface of the Earth to the Moon as from the surface of the Earth to the centre of the Earth.

In modern terms, the Moon is 381,000 kilometres (237,000 miles) from Earth and has a diameter of 2,470 kilometres (2,160 miles).

The Greeks already knew that the Moon was the nearest of the heavenly bodies and that the other objects were all much farther away. To be so much farther away and to be visible at all, they must all be worlds in size.

The notion of the plurality of worlds descended from the rarefied heights of philosophic speculation to the literary level with the first account we know of that reads like modern science fiction stories involving interplanetary travel.

About AD 165, a Greek writer named Lucian of Samosata wrote *A True History,* an account of a trip to the Moon. In that book, the hero is carried to the Moon by a whirlwind.

He finds the Moon luminous and shining, and in the distance he can see other luminous worlds. Down below, he sees a world that is clearly his own world, the Earth.

Lucian's universe was behind the scientific knowledge of his own time, since he had the Moon glowing and he had the heavenly bodies all close together. Lucian also assumed that air filled all of space and that 'up' and 'down' were the same everywhere. There was no reason as yet to think that that was not so.

Every world in Lucian's universe was inhabited, and he assumed the presence of extraterrestrial intelligence everywhere. The king of the Moon was Endymion and he was at war with the king of the Sun, Phaethon. (These names were taken out of the Greek myths, where Endymion was a youth beloved by the Moon goddess, and Phaethon was the son of the Sun god.) The Moon beings and Sun beings were quite human in appearance, in institutions, and even in their follies, for Endymion and Phaethon were at war with each other, disputing the colonization of Jupiter.

It was not for nearly 1,300 years, however, that a major writer dealt with the Moon again. This came in 1532 in *Orlando Furioso*, an epic poem written by the Italian poet Ludovico Ariosto (1474–1533). In it, one of the characters travels to the Moon in the divine chariot that carried the prophet Elijah in a whirlwind to Heaven. He finds the Moon well populated by civilized people.

The notion of a plurality of worlds received still another push forwards with the invention of the telescope. In 1609, the Italian scientist Galileo Galilei (1564–1642) constructed a telescope and pointed it at the Moon. For the first time in history, the Moon was seen magnified, and more clearly detailed than was possible with the unaided eye.

Galileo saw mountain ranges on the Moon, together with what looked like volcanic craters. He saw dark, smooth patches that looked like seas. Quite plainly and simply, he was seeing another world.

This stimulated the further production of fictional flights to the Moon. The first was written by Johannes Kepler (1571–1630), an astronomer of the first rank, and was published posthumously in 1633. It was entitled *Somnium* because the hero reached the Moon in a dream.*

The book was remarkable in that it was the first to take into account the actual known facts about the Moon, which until then had been treated as in no way different from any Earthly piece of real estate. Kepler was aware that on the Moon the nights and days were each fourteen Earth days long. However, he had air, water, and life on the Moon; there was nothing as yet to rule that out.

In 1638, the first science fiction story in the English language that dealt with a flight to the Moon was published. It was *The Man in the Moone* by an English bishop named Francis Godwin (1562–1633). It was also published posthumously.

Godwin's book was the most influential of the early books of this nature, for it inspired a number of imitations. The hero of the book was carried to the Moon in a chariot drawn by a flock of geese (who were pictured as regularly migrating to the Moon). As usual, the Moon was populated with quite human intelligent beings.

In the same year in which Godwin's book was published, another English bishop, John Wilkins (1614–1672), a brother-in-law of Oliver Cromwell, produced a nonfictional equivalent. In his book *The Discovery of a World in the Moone*, he speculated on the habitability of that body. Whereas Godwin's hero was a Spaniard (the Spaniards having been the great explorers of the previous century), Wilkins was sure it would be an Englishman who would first reach the Moon. In a way, Wilkins proved right, for the first man on the Moon was of English descent.

Wilkins, too, assumed that air existed all the way to the Moon and indeed throughout the Universe. There was, even in 1638, no understanding that such a fact would make separate heavenly bodies impossible. If the Moon were revolving

* It was the first science fiction story to be written by a professional scientist – but not, by a long shot, the last.

about the Earth through an infinite ocean of air, air resistance would gradually slow it and finally bring it crashing, in fragments, down on the Earth – which would similarly crash into the Sun, and so on.

Waterlessness

The notion of universal air had not long to live, however. In 1643, the Italian physicist Evangelista Torricelli (1608–1647), a student of Galileo, succeeded in balancing the weight of the atmosphere against a column of mercury, inventing the barometer. It turned out, from the weight of the column of mercury that balanced the downward pressure of air, that the atmosphere would only be 8 kilometres (5 miles) high if it were of uniform density. And if the density decreased with height, as it does in fact, it could only be a little higher than that before becoming too thin to support life.

It was clear, for the first time, that air did not fill the Universe but was a purely local terrestrial phenomenon. The space between the heavenly bodies was empty, a 'vacuum', and this constituted, in a way, the discovery of outer space.

Without air, human beings could not travel to the Moon by means of water spouts, or geese-drawn chariots, or by any of the usual methods that would suffice to cross a gap of air.

The only way, in fact, that the gap between Earth and Moon could be closed was by using rockets, and this was first mentioned in 1647 by none other than the French writer and duellist Savinien de Cyrano de Bergerac (1619–1655). Cyrano, in his book *Voyages to the Moon and the Sun*, listed seven different ways in which a human being might travel from the Earth to the Moon, and one of them was by means of rockets. His hero actually performed the voyage, however, by one of the other (alas, worthless) methods.

As the seventeenth century progressed and as observation of the Moon continued with better and better telescopes, astronomers grew aware of certain peculiarities about our satellite.

The view of the Moon, it seemed, was always clear and un-changing. Its surface was never obscured by cloud or mist. The terminator – that is, the dividing line between the light and the dark hemispheres – was always sharp. It was never fuzzy as it would be if light were refracting through an atmo-sphere, thus signifying the presence on the Moon of the equivalent of an Earthly twilight.

What's more, when the Moon's globe approached a star, the star remained perfectly bright until the Moon's surface reached it and then it winked out in an instant. It did not slowly dim as it would if the Moon's atmosphere reached it before the Moon's surface did, and if the starlight had to penetrate thickening layers of air.

In short, it became clear that the Moon was an airless world. And waterless, too, for closer examination showed that the dark 'seas' that Galileo had seen were speckled with craters here and there. They were, if anything, seas of sand, but cer-tainly not of water.

Without water, there could scarcely be life on the Moon. For the first time, people had to become aware that it was possible for a dead world to exist; one that was empty of life.

Let us not, however, hasten too quickly. Given a world with-out air and water, can we be sure it has no life?

Let us begin by considering life on Earth. Certainly, it shows a profound variability and versatility. There is life in the ocean deeps and on the ocean surface, in fresh water and on land, underground, in the air, even in deserts and frozen wastes.

There are even microscopic forms of life that do not use oxygen and to some of which oxygen is actually deadly. For them, airlessness would have no fears. (It is because of them that food sealed in a vacuum must be well heated first. Some pretty dangerous germs, including the one that produces botulism, get along fine in a vacuum.)

Well, then, is it so difficult to imagine some forms of life getting along without water, too?

Yes, quite difficult. No form of terrestrial life can do without water. Life developed in the sea, and the fluids within the living cells of all organisms, even those who now live in fresh water or on dry land and who would die if placed in the sea, are essentially a form of ocean water.

Even the life forms in the driest desert have not evolved into independence of water. Some might never drink, but they then get their necessary water in other ways – from the fluids of the food they eat, for instance – and carefully conserve what they get.

Some bacteria can survive desiccation and, in spore form, can live on for an indefinite period without water. The spore wall, however, protects the fluid within the bacterial cell. True desiccation, through and through, would kill it as quickly as it would kill us.

Viruses can retain the potentiality of life even when crystallized and with no water present. They cannot multiply, however, until they are within a cell and can undergo changes within the milieu of the cell fluid.

Ah, but all this refers to Earth life, which has developed in the ocean. On a waterless world, might not a fundamentally different kind of life develop that *was* independent of water?

Let's reason this out as follows:

On the surface of planetary worlds (on one of which the one example of life that we know of has developed) matter can exist in any of three stages: solid, liquid, or gas.

In gases, the component molecules are separated by relatively large distances and move randomly. For that reason, gas mixtures are always homogeneous, that is, all components are well mixed. Any chemical reaction that takes place in one part can equally well take place in another part and therefore spreads from one part of the system to the other with explosive rapidity. It is difficult to see how the carefully controlled and regulated reactions, which seem essential to something that is as complicated and finely balanced as living systems would appear to be, can exist in a gas.

Then, too, the molecules making up gases tend to be very simple. The complicated molecules that we can assume would be needed (if we are expected to witness the varied, versatile, and subtle changes that must surely characterize anything as varied, versatile, and subtle as life) are, under ordinary circumstances, in the solid state.

Some solids can be converted into gases by being heated sufficiently, or by being put under very low pressure. The complicated molecules characteristic of life would break up into small fragments if heated, however, and would be useless. If placed under even zero pressure, the complicated molecules will produce only insignificant quantities of vapour.

We conclude, then, that we cannot have life in the gaseous state.

In solids, the component molecules are in virtual contact, and can exist to any degree of complication. What's more, solids can be, and usually are, heterogeneous; that is, the chemical makeup in one part can be quite different from the chemical makeup in another part. In other words, different reactions can take place in different places at different rates and under different conditions.

So far, so good, but the trouble is that the molecules in solids are more or less locked in place, and chemical reactions will take place too slowly to produce the delicate changeability we associate with life. We conclude, then, we cannot have life in the solid state.

In the liquid state, the component molecules are in virtual contact, and the possibility of heterogeneity exists, as in the solid state. However, the component molecules move about freely, and chemical reactions can proceed quickly, as in the gaseous state. What's more, both solid and gaseous substances can dissolve in liquids to produce extraordinarily complicated systems in which there is no limit to versatility of reaction.

In short, the kind of chemistry we associate with life would seem to be possible only against a liquid background. In Earth's case that liquid is water, and we will have something to say later in the book as to whether there is a possibility of any substitute.

A world, then, that is without water (and without any other

liquid that might substitute) would seem to be surely incapable of supporting life.

Or am I still being too narrow minded?

Why can't life, with chemical and physical properties completely different from terrestrial life, nevertheless develop and even evolve intelligence? Why can't there be a very slow, solid life form (too slow, perhaps, to be recognized as life by us) living on the Moon or, for that matter, here on Earth? Why not a very rapid and evanescent gaseous life form, literally exploding with thought and experiencing lifetimes in split seconds, existing on the Sun, for instance.

There have been speculations in this direction. Science fiction stories have been written that postulated enormously strange life forms. The Earth itself has been considered as a living being, as have whole galaxies, and as have clouds of dust and gas in interstellar space. Life consisting of pure energy radiation has been written about and life existing outside our Universe altogether and therefore indescribable.

There is no limit to speculation in this respect, but in the absence of any evidence, they can only remain speculations. In this book, however, I will move only in those directions in which there is at least some evidence to guide me. Fragmentary and tenuous that evidence may be, and the conclusions shaky enough – but to step across the line into the region of no evidence at all I will not do.

Therefore, *until evidence to the contrary is forthcoming*, I must conclude that, on the basis of what we know of life (admittedly limited), a world without liquid is a world without life. Insofar then as the Moon seems to be a world without liquid, the Moon would seem to be a world without life.

We might be more cautious and say that a world without liquid is a world without life-as-we-know-it. It would be tiresome, however, to repeat the phrase constantly, and I will say it only now and then to make sure you don't forget that that is what I mean. In between, please take it for granted that in this book I am speaking of life-as-we-know-it, whenever I speak of life. Please remember also that there is not one scrap of evidence, however faint or indirect, that speaks for the existence of life-not-as-we-know-it.

Even now, we may be rushing to a conclusion too rapidly. The astronomers at their early telescopes could see clearly that there was no water on the Moon in the sense that there were no seas, great lakes, or mighty rivers. As telescopes continued to improve, no sign of 'free water' on the surface ever showed up.

Yet might there not be water present in minor quantities, in small pools or bogs in the shadow of crater walls, in underground rivers and seepages, or even just in loose chemical combination with the molecules making up the Moon's solid surface?

Such water would surely not be observable through a telescope, and yet it might be enough to support life.

Yes, it might – but if life had its origin through chemical reactions taking place randomly (and we will discuss this in a later chapter), then the larger the volume in which those random processes take place, the greater the chance that they would finally succeed in producing something as complicated as life. Furthermore, the larger the volume in which the process took place, the more room there would be for the kind of prodigal outpouring of death and replacement that serves as the power drive for the random process of evolution.

Where only small quantities of water exist, the formation of life becomes very unlikely; and if it does form, its evolution is very slow. It simply passes the bounds of likelihood that there would be time and opportunity for a complex life form to arise and flourish, certainly not one complex enough to develop intelligence and a technological civilization.

Consequently, even if we admit the presence of water in quantities not visible through the telescope, we can at best postulate only very simple life. There is no way in which we can imagine the Moon to be the home of extraterrestrial intelligence – assuming it has always been as it is now.

Moon Hoax

Again I say that it is *not* the concept of extraterrestrial intelligence that is hard to grasp. It is the reverse notion that meets with resistance. Telescopic evidence (in the Moon's case) to the contrary, it remained hard to imagine dead worlds.

In 1686, the French writer Bernard le Bovier de Fontenelle (1657–1757) wrote *Conversations on the Plurality of the Worlds*, in which he speculated charmingly on life on each of the then known planets from Mercury to Saturn.

And though the case of life on the Moon was already dubious in Fontenelle's time and grew steadily more dubious, it proved quite possible to hoodwink the general public with tales of intelligent life on the Moon as late as 1835. That was the year of the 'Moon Hoax'.

This took place in the columns of a newly established newspaper, the *New York Sun*, which was eager to attract attention and win readers. It hired Richard Adams Locke (1800–1871), an author who had arrived in the United States three years before from his native England, to write essays for them.

Locke was interested in the possibility of life on other worlds and had even tried his hand at science fiction in that connection. Now it occurred to him to write a little science fiction without actually saying that that was what it was.

He chose for his subject the expedition of the English astronomer John Herschel (1792–1871). Herschel had gone to Cape Town in southern Africa to study the southern sky.

Herschel had taken good telescopes with him, but they were not the best in the world. Their value lay not in themselves but in the fact that since all astronomers and astronomical observatories were at that time located in the northern hemisphere, the regions near the South Celestial Pole had virtually never been studied at all. Almost any telescope would have been useful.

Locke knew well how to improve on that. Beginning with the 25 August 1835 issue of the *Sun*, Locke carefully described all sorts of impossible discoveries being made by Herschel with a telescope capable (so Locke said) of such magnification that it could see objects on the Moon's surface that were only eighteen inches across.

In the second day's instalment, the surface of the Moon was described. Herschel was said to have seen flowers like poppies and trees like yews and firs. A large lake, with blue water and foaming waves, was described, as were large animals resembling bisons and unicorns.

One clever note was the description of a fleshy flap across the forehead of the bisonlike creatures, a flap that could be raised or lowered to protect the animal 'from the great extremes of light and darkness to which all the inhabitants of our side of the Moon are periodically subject'.

Finally, creatures with human appearance, except for the possession of wings, were described. They seemed to be engaged in conversation: 'their gesticulation, more particularly the varied action of their hands and arms, appeared impassioned and emphatic. We hence inferred that they were rational beings.'

Astronomers, of course, recognized the story to be nonsense, since no telescope then built (or now, either) could see such detail from the surface of the Earth, and since what was described was utterly at odds with what was known about the surface of the Moon and its properties.

The hoax was revealed as such soon enough, but in the interval the circulation of the *Sun* soared until, for a brief moment, it was the best-selling newspaper in the world. Uncounted thousands of people believed the hoax implicitly and remained eager for more, showing how anxious people were to believe in the matter of extraterrestrial intelligence – and indeed in any dramatic discovery (or purported discovery) that seems to go against the rational but undramatic beliefs of realistic science.

As the Moon's deadness became more and more apparent, however, hope remained that this was an unusual and an isolated case; and that the other worlds of the Solar System might be inhabited.

When the English mathematician William Whewell (1794–1866), in his book *Plurality of Worlds* published in 1853, suggested that some of the planets might not bear life, this definitely represented a minority opinion at the time. In 1862, the young French astronomer Camille Flammarion (1842–1925) wrote *On the Plurality of Habitable Worlds* in refutation, and this second book proved much the more popular.

Soon after the appearance of Flammarion's book, however, a new scientific advance placed the odds heavily in Whewell's favour.

Airlessness

In the 1860s, the Scottish mathematician James Clerk Maxwell (1831–1879) and the Austrian physicist Ludwig Edward Boltzmann (1844–1906), working independently, advanced what is called the kinetic theory of gases.

The theory considered gases as collections of widely spaced molecules moving in random directions and in a broad range of speeds. It showed how the observed behaviour of gases under changing conditions of temperature and pressure could be deduced from this.

One of the consequences of the theory was to show that the average speed of the molecules varied directly with the absolute temperature, and inversely with the square root of the mass of the molecules.

A certain fraction of the molecules of any gas would be moving at speeds greater than the average for that temperature, and might exceed the escape velocity for the planet whose gravitational attraction held them. Anything moving at more than escape velocity, whether it is a rocket ship or a molecule, can, if it does not collide with something, move away for ever from the planet.

Under ordinary circumstances, so tiny a fraction of the molecules of an atmosphere might attain escape velocity – and retain it through inevitable collisions until it reached such heights that it could move away without further collision – that the atmosphere would leak away into outer space with imperceptible slowness. Thus, Earth, for which the escape velocity is 11.3 kilometres (7.0 miles) per second, holds on to its atmosphere successfully and will not lose any significant quantity of it for billions of years.

If, however, Earth's average temperature were to be substantially increased, the average speed of the molecules in its atmosphere would also be increased and so would the fraction of those molecules travelling at more than escape velocity. The atmosphere would leak away more rapidly. If the temperature were high enough, the Earth would lose its atmosphere rather quickly and become an airless globe.

Next, consider hydrogen and helium, which are gases that are composed of particles much less massive than those making up the oxygen and nitrogen of our atmosphere. The oxygen molecule (made up of 2 oxygen atoms) has a mass of 32 in atomic mass units, and the nitrogen molecule (made up of 2 nitrogen atoms) has a mass of 28. In contrast, the hydrogen molecule (made up of 2 hydrogen atoms) has a mass of 2 and helium atoms (which occur singly) a mass of 4.

At a given temperature, light particles move more rapidly than massive ones. A helium atom will move about three times as quickly as the massive and therefore more sluggish molecules of our atmosphere, and a hydrogen molecule will move four times as quickly. The percentage of helium atoms and hydrogen molecules that would be moving more rapidly than escape velocity would be much greater than in the case of oxygen and nitrogen.

The result is that Earth's gravity, which suffices to hold the oxygen and nitrogen molecules of its atmosphere indefinitely, would quickly lose any hydrogen or helium in its atmosphere. That would leak away into outer space. If the Earth were forming under its present condition of temperature and were surrounded by cosmic clouds of hydrogen and helium, it would not have a sufficiently strong gravitational field to collect those small and nimble molecules and atoms.

It is for this reason that Earth's atmosphere does not contain anything more than traces of hydrogen and helium, although these two gases make up by far the bulk of the original cloud of material out of which the Solar System was formed.

The Moon has a mass only $\frac{1}{81}$ that of the Earth and a gravitational field only $\frac{1}{81}$ as intense. Because it is a smaller body than the Earth, its surface is nearer its centre, so that its small gravitational field is somewhat more intense at its surface than you would expect from its overall mass. At the surface, the Moon's gravitational pull is $\frac{1}{6}$ of the Earth's gravitational pull at its surface.

This is reflected in escape velocity as well. The Moon's escape velocity is only 2.37 kilometres (1.47 miles) per second. On Earth, a vanishingly small percentage of molecules of a particular gas might surpass its escape velocity. On the Moon, a substantial percentage of molecules of that same gas would surpass the Moon's much lower escape velocity.

Then, too, because the Moon rotates on its axis so slowly as to allow the Sun to remain in the sky over some particular point on its surface for two weeks at a time, its temperature during its day rises much higher than does the Earth's temperature. That further increases the percentage of molecules with speeds surpassing the escape velocity.

The result is that the Moon is without an atmosphere. To be sure, even the Moon's low gravity can hold some gases if their atoms or molecules are massive enough. The atoms of the gas krypton, for instance, have a mass of 83.8 and the atoms of the gas xenon, a mass of 131.3. The Moon's gravitational field could hold them with ease. However, these gases are so uncommon in the Universe generally, that even if they occurred on the Moon and made up its atmosphere, that atmosphere would be only a trillionth as dense as the Earth's atmosphere, if that, and could at best be described as a 'trace atmosphere'.

To all intents and purposes, as far as the problem of

extraterrestrial life is concerned, such a trace atmosphere is of no consequence and the Moon can still fairly be described as airless.

All this has meaning with respect to a liquid such as water. Water is 'volatile', that is, it has a tendency to vaporize and turn into a gas. At a given temperature, there is a counter-tendency for the gaseous water vapour to recondense into liquid. At any particular temperature, liquid water is therefore liable to be in equilibrium with a certain pressure of water vapour, provided that water vapour is not removed from the vicinity as, for instance, by a wind.

If the water vapour is removed, equilibrium pressure is not built up and more of the liquid water vaporizes, and still more, till it is all gone. We are all familiar with the way in which the water left behind by a rainstorm evaporates until it is finally all gone. The higher the temperature, the faster the water evaporates.

Naturally, the water vapour is not removed from the Earth altogether. If it does not condense in one place, it condenses in another as dew, fog, rain, or snow, and thus the Earth holds on to its water.

If there were liquid water on the Moon, the vapour that would form would leak out into space, for the mass of the water molecule is but 18 and the Moon's gravitational field would not hold it. The liquid water would continue to vaporize and eventually the Moon would dry up altogether. The fact that there is no air on the Moon means there is no air pressure to slow the rate of water evaporation, and the water, if it had been present, would have been lost all the more quickly.

The Moon, therefore, *must* be without water as well as without air. What's more, any airless world would be a lifeless world – not because air is necessarily essential to life, but because an airless world is a waterless world, and water is essential.

Even the kinetic theory of gases leaves loopholes, however. The possibility remains that scraps of water, even air, can exist underground on the Moon, or in chemical combination with molecules in the soil. In that case, the small molecules would be prevented from leaving by forces other than gravity – by physical barriers or chemical bonding.

Then, too, there may have been a time early in the history of the Moon when it had an atmosphere and an ocean, *before* it lost them both to space. Perhaps in those early days, life developed, even intelligent life, and it may have adapted itself, either biologically or technologically, to the gradual loss of air and water. It might, therefore, be living on the Moon in caverns, with a supply of air and water sealed in.

As late as 1901, the English writer H. G. Wells (1866–1946) could publish *The First Men on the Moon* and have his heroes find a race of intelligent Moon beings, rather insectlike in character and highly specialized, living underground.

Even that much seems doubtful, however, since calculations show that the Moon would have lost its air and water (if any) quite rapidly. It would have retained them for many times the lifetime of a human being, of course, and if we were living on the Moon when it still had an atmosphere and ocean we could live out our life normally. The atmosphere and ocean would not last long enough, however, to allow life to develop and intelligence to evolve from zero. It wouldn't even come close to doing that.

And we seem to be at a final answer now. On 20 July 1969, the first astronauts landed on the Moon. Samples of material from the Moon's surface were brought back on this and later trips to the Moon. Apparently the Moon rocks all seem to indicate that the Moon is bone dry; that there is no trace of water upon it, nor has there been in the past.

The Moon would seem to be, almost beyond conceivable doubt, a dead world.

3 The inner Solar System

Nearby worlds

Once Galileo began to study the sky with his telescope, he could see that the various planets expanded into tiny orbs. They appeared as mere dots of light to the unaided eye merely because of their great distance.

What's more, Venus, being closer to the Sun than Earth is, showed phases like the Moon, as it should under such conditions if it were a dark body shining only by reflection. That was proof enough that the planets were also worlds, possibly more or less Earthlike.

Once that was established, it was taken for granted that all of them were life bearing and inhabited by intelligent creatures. Flammarion maintained this confidently, as I said in the previous chapter, as late as 1862.

The kinetic theory of gases, however, ruled out not merely the Moon as an abode of life, but any world smaller than itself. Any worlds smaller than the Moon could scarcely be expected to possess air or water. They would lack the gravitational field for it. Consider the asteroids, the first of which was discovered in 1801. They circle the Sun just outside the orbit of Mars and the largest of them is but 1,000 kilometres (620 miles) in diameter. There are anywhere from 40,000 to 100,000 of them with diameters of at least a kilometre or two, and every last one of them lacks air or liquid water and are therefore without life.*

---

* There may be small amounts of water in the solid state (ice) held to the asteroids and other small worlds by chemical bonds that don't depend on gravitational forces for their efficacy. Frozen water, however, is not suitable for life and even on Earth the frozen ice sheets of Greenland and Antarctica are life free in their natural state.

The same is true for the two tiny satellites of Mars, discovered in 1877. They are in all likelihood captured asteroids, and have neither air nor liquid water.

Within the orbits of the asteroids lies the 'inner Solar System' and there we find four planetary bodies larger than the Moon. In addition to the Earth itself, we have Mercury, Venus, and Mars.

Of these, Mercury is the smallest, but it is 4.4 times as massive as the Moon and its diameter is 4,860 kilometres (3,020 miles), which is 1.4 times that of the Moon. Mercury's surface gravity is 2.3 times that of the Moon and nearly $\frac{2}{5}$ that of the Earth. Might it not manage to retain a thin atmosphere?

Not so. Mercury is also the closest of the planets to the Sun. At its nearest approach to the Sun it is at only $\frac{3}{10}$ the distance from it that the Earth is. Any air it might have would be heated to far higher temperatures than the Earth's atmosphere. Gas molecules on Mercury would be correspondingly speedier in their motion and harder to hold on to. Mercury, therefore, would be expected to be as airless and waterless – and as lifeless – as the Moon.

In 1974 and 1975, a rocket probe, *Mariner 10*, passed near Mercury's surface on three occasions. On the third occasion, it passed within 327 kilometres (203 miles) of the surface. Mercury was mapped in detail and its surface was found to be cratered in a very Moonlike way, and its airlessness and waterlessness was confirmed. There is no perceptible doubt as to its lifelessness.

Venus looks far more hopeful. Venus's diameter is 12,100 kilometres (7,520 miles) as compared with Earth's 12,740 kilometres (7,920 miles). Venus's mass is about 0.815 times that of the Earth and its surface gravity is 0.90 times that of the Earth.

Even allowing for the fact that Venus is closer to the Sun than Earth and would therefore be hotter than Earth, it would seem that Venus should have an atmosphere. Its gravitational field is strong enough for that.

And, indeed, Venus *does* have an atmosphere, a very pronounced one, and one that is far cloudier than ours. Venus is wrapped in a planet-girdling perpetual cloud cover, which was at once taken as adequate evidence that there was water on Venus.

The cloud cover does, unfortunately, detract from the hopeful views we can have of Venus, since it prevents us from gathering evidence as to its fitness for life. At no time could astronomers ever catch a glimpse of its surface, however good their telescopes. They could not tell how rapidly Venus might rotate on its axis, how tipped that axis might be, how extensive its oceans (if any) might be, or anything else about it. Without more evidence than the mere existence of an atmosphere and clouds it was difficult to come to reasonable conclusions about life on Venus.

Mars's on the other hand, is at once less hopeful and more hopeful.

It is less hopeful because it is distinctly smaller than Earth. Its diameter is only 6,790 kilometres (4,220 miles) and its mass is only 0.107 that of the Earth. With a mass only $\frac{1}{10}$ that of Earth it is not exactly a large world, but on the other hand it is 8.6 times as massive as the Moon, so it is not exactly a small one, either. It is, in fact, twice as massive as Mercury.

Mars's surface gravity is 2.27 times that of the Moon and is just about that of Mercury. Mars, however, is four times as far from the Sun as Mercury is, so that Mars is considerably the cooler of the two. Mars's gravitational field need deal with considerably slower molecules for that reason.

It follows that although Mercury is without an atmosphere, Mars may have one – and it does. Mars's atmosphere is a thin one, to be sure, but it is distinctly there. Mars is presumably drier than the Earth, for its atmosphere is not as cloudy as Earth's (let alone Venus's), but occasional clouds are seen. Dust storms are also seen, so there must be sharp winds on Mars.

The more hopeful aspect of Mars is that its atmosphere is sufficiently thin and cloud free to allow its surface to be seen (rather vaguely) from Earth. For centuries, astronomers have done their best to map what it was they saw on that distant world. (At its closest, Mars can approach as closely as

56,000,000 kilometres [34,800,000 miles] to Earth, a distance that is 146 times as far away from us as the Moon.)

The first to make out a marking that others could see as well was the Dutch astronomer Christiaan Huygens (1629–1695). In 1659, he followed the markings he could see as they moved around the planet and determined the rotation period of Mars to be only a trifle longer than that of Earth. We now know Mars rotates in 24.66 hours compared to Earth's twenty-four.

In 1781, the German-English astronomer William Herschel (1738–1822) noted that Mars's axis of rotation was tilted to the perpendicular, as Earth's was, and almost by the same amount. Mars's axial tilt is 25.17° as compared with Earth's 23.45°.*

This means that not only does Mars have a day-night altera-tion much as Earth has, but also seasons. Of course, Mars is half again as far from the Sun as we are, so that its seasons are colder than ours. Furthermore, it takes Mars longer to com-plete its orbit about the Sun, 687 days to our 365¼, so that the seasons on Mars are each nearly twice as long as ours.

In 1784, Herschel noted that there were ice caps about the Martian poles, as there were about Earth's poles. There was one more point of resemblance in that the ice caps were as-sumed to be frozen water, and therefore to serve as proof there was water on Mars.

Mars and Venus both looked like hopeful possible abodes of life, certainly far more hopeful than the asteroids or the Moon or Mercury.

Venus

In 1796, the French astronomer Pierre Simon de Laplace (1749–1827) speculated on the origin of the Solar System.

The Sun rotates on its axis in an anticlockwise direction when viewed from a point far above its north pole. From that same point, all the planets known to Laplace moved about the

---

* William Herschel was the father of John Herschel, who a half-century later was to be victimized by the Moon Hoax.

Sun in an anticlockwise direction, and all the planets whose rotations were known rotated about their axes in an anticlockwise direction. Added to that was the fact that all the satellites known to Laplace revolved about their planets in an anticlockwise direction.

Finally, all the planets had orbits that were nearly in the plane of the Sun's equator and all the satellites had orbits that were nearly in the plane of their planet's equator.

To account for all this, Laplace suggested that the Solar System was originally a vast cloud of dust and gas called a nebula (from the Latin word for *cloud*). The nebula was turning slowly in an anticlockwise direction. Its own gravitational field slowly contracted it, and as it contracted it had to spin faster and faster in accordance with something called the law of conservation of angular momentum. Eventually, it condensed to form the Sun, which is still spinning in the anticlockwise direction.

As the cloud contracted on its way to the Sun and as it increased its rate of spin, the centrifugal effect of rotation caused it to belly out at its equator. (This happens to the Earth, which has an equatorial bulge that lifts points on its equator thirteen miles farther from the centre of the Earth than the north and south poles are.)

The bulge of the contracting nebula became more and more pronounced as it shrank further and speeded up further, until the entire bulge was thrown off like a thin doughnut around the contracting nebula. As the nebula continued to shrink, additional doughnuts of matter were shed.

Each doughnut, in Laplace's view, gradually condensed into a planet, maintaining the original anticlockwise spin, and speeding up that spin as it condensed. As each planet formed there was a chance it might shed smaller subsidiary doughnuts of its own, which became the satellites. The rings around Saturn are examples of matter that has been given off (according to Laplace's nebular hypothesis) and has not yet condensed to a satellite.

The nebular hypothesis explains why all the revolutions and

rotations in the Solar System should be in the same direction.*
It is because all participate in the spin of the original nebula.

It also explains why all the planets revolve in the plane of
the Sun's equator. It is because it is from the Sun's equatorial
regions that they were originally formed; and it is from the
planetary equatorial regions that the satellites formed.

The nebular hypothesis was more or less accepted by
astronomers during the nineteenth century, and it added detail
to the picture that people drew of Mars and Venus.

As the nebula condensed, according to this theory, it would
seem that the planets would form in order from the outermost
to the innermost. In other words, after the nebula had condensed to the point where it was only 500,000,000 kilometres
(310,000,000 miles) across, it gave off the ring of matter that
formed Mars. Then, after considerable time taken up in further
contraction, it gave off the matter that formed the Earth and
the Moon, and after another unknown length of time, the
matter that formed Venus.

By the nebular hypothesis, therefore, it would seem that
Mars was considerably older than Earth, and that Earth was
considerably older than Venus.

It became customary, therefore, to think of Mars as having
moved farther along the evolutionary path than Earth; not
only with respect to its planetary characteristics, but with respect to the life upon it. Similarly, Venus had not moved as
far along the evolutionary path. Thus, the Swedish chemist
Svante August Arrhenius (1859–1927) drew an eloquent picture in 1918 of Venus as a water-soaked jungle.

This sort of thinking was reflected in science fiction stories,
which very often depicted Mars as occupied by an intelligent
race with a long history that dwarfed that of Earthly human
beings. The Martians were pictured as far advanced beyond
us technologically, but often as decadent and weary of life –
in their old age as a species.

On the other hand, many stories were written of a jungle-
like Venus, or one with a planetary ocean – in either case filled

* Today, we know of some exceptions.

with primitive life forms. In 1954, I myself published a novel, *Lucky Starr and the Oceans of Venus*, in which the planet was described as having a planetary ocean. But only two years later our thoughts about Venus were revolutionized.

After World War II, astronomers gained a large number of new and extraordinarily useful tools for the exploration of the worlds of the Solar System. They could send out microwaves to the surfaces of distant planets, receive the reflections, and from the properties of those reflections deduce the nature of the surface even if they could not see them optically. They could receive radio waves sent out by the planets themselves. They could send out rocket-powered probes that could skim by the planet or even land on their surfaces and send back useful data (as in the case of the mapping of Mercury's surface by *Mariner 10*).

In 1956, the American astronomer Robert S. Richardson analysed radar reflections from Venus's surface beneath the cloud layer and found it was rotating, very slowly, in the wrong direction – clockwise.

In the same year, a team of astronomers under Cornell H. Mayer received radio waves from Venus and were astonished to find that the intensity of those waves was equivalent to what would be expected from an object far hotter than Venus was thought to be. If this were so, there could be no planetary ocean on Venus; indeed no liquid water of any kind (and there went my poor novel when it was only two years old).

On 14 December 1962, an American Venus probe, *Mariner 2*, passed close by Venus's position in space, monitored its radio-wave emission, and confirmed the earlier report. On 12 June 1967, a Soviet Venus probe, *Venera 4*, entered Venus's atmosphere and sent back confirming data while descending for an hour and a half. *Venera 5* and 6, landing on Venus's surface on 16 and 17 May 1969, put the matter beyond all doubt.

Venus has an extraordinarily dense atmosphere, about ninety-five times as dense as Earth's. Venus's atmosphere, what's more, is 95 per cent carbon dioxide, the molecules of which have a mass of 44. (Carbon dioxide had been detected in Venus's atmosphere by more ordinary methods as long before as 1932.)

It is natural enough for a planet to have an atmosphere containing carbon dioxide. Our own atmosphere has a small quantity of carbon dioxide (0.03 per cent) and that small quantity is essential to the growth of plant life.

The photosynthesis of green plants uses the energy of the Sun to combine carbon dioxide molecules with water molecules to form the components of plant tissue – sugar, starch, cellulose, fats, proteins, and so on. In the process, though, free oxygen is formed in excess and is discharged into the atmosphere.

It is generally thought, in fact, that at some time in the distant past, the Earth's atmosphere was far richer in carbon dioxide that it is now, and that free oxygen was absent. (We'll get back to this matter later in the book.) Earth's early atmosphere, then, was somewhat like Venus's present one, but less dense; and it is only the action of photosynthesis that gradually removed the carbon dioxide and replaced it with oxygen.

From the fact that Venus's atmosphere is so rich in carbon dioxide and so poor in oxygen (none has been detected), we can deduce at once that photosynthesis in its Earthly form is absent from the planet or, at the very least, has not been established for long.

This would seem to indicate that there are no green plants of any consequence on the planet, and therefore no animal life (which depends ultimately on plants for food), and therefore no intelligence.

It might be argued that photosynthesis is not essential to life and, indeed, it isn't. On Earth there are forms of life that neither use photosynthesis nor depend on other forms of life that use photosynthesis. These forms of life are all at the bacterial level, however, and there is no indication that now or ever has any form of life beyond the bacterial existed on Earth without need, direct or indirect, of photosynthesis.

It might also be argued that Earth need not form a rule in

this respect. Suppose a form of life got its energy from the Sun and made use of carbon dioxide, but somehow stored the oxygen instead of emitting it into the atmosphere. In due course of time, it made use of the oxygen for the purpose of combining it with carbon atoms and restoring carbon dioxide to the atmosphere. In that way, you could have photosynthesis while retaining a carbon dioxide atmosphere.

This is not beyond the bounds of possibility, but ...

Carbon dioxide has the property of absorbing infrared radiation. It allows the high-energy visible light of the Sun to pass through and strike the surface of a planet, but then absorbs the low-energy (and invisible) infrared radiation the planet reemits to space at night. This is called the greenhouse effect because the glass of a greenhouse does the same thing.

By retaining the infrared radiation of the planet, the carbon dioxide in the atmosphere raises the temperature of the planet, as the glass's retention of infrared radiation raises the temperature inside a greenhouse. Because of the very high content of carbon dioxide in Venus's atmosphere, the surface temperature of the planet is far higher than we would expect it to be from its distance from the Sun alone, especially since ordinarily we would expect its clouds to shield it from much of the Sun's heat. Venus is the victim of a runaway greenhouse effect.

The result is that Venus's surface temperature is about $480°C$ ($900°F$), considerably higher than Mercury's surface temperature. Mercury may be closer to the Sun, but it doesn't have a heat-conserving atmosphere.

The surface temperature of Venus is far above the boiling point of water and is, indeed, hot enough to melt lead easily. There can be no liquid water anywhere on the planet. What water it has must exist as vapour in the clouds, and there is evidence that the liquid droplets in the clouds are, to a considerable extent, the extremely corrosive substance sulphuric acid.

It takes a vivid imagination indeed to conceive of life on such a planet, and Venus must be crossed off as a possible abode for extraterrestrial intelligence.

Martian canals

As for Mars, that from the beginning seemed to have a much better chance for life. Its rotation, its axial tip, its ice caps all seemed hopeful. Its presumed great age gave it, it would seem, a particularly good chance of advanced life.

About 1830, astronomers began to make serious attempts to map Mars. The first map produced was by a German astronomer, Wilhelm Beer (1797–1850). Others followed, but success was not remarkable. It was hard to see details through two atmospheres, those of Earth and of Mars, from a distance of hundreds of millions of kilometres. Each astronomer who tried to map Mars seemed to end up with a map that was completely unlike the ones produced by his predecessors.

All agreed, however, that there seemed to be light areas and dark areas, and the notion grew that the light areas represented land surface and the dark areas water surface.

A particularly good chance for observation came in 1877 when Mars and Earth happened to be in those parts of their orbits that brought them as closely together as they ever got to be. And by then, of course, astronomers had better telescopes than they ever had before.

One observer with an excellent telescope was the Italian astronomer Giovanni Virginio Schiaparelli (1835–1910). During his observations in 1877, he drew a map of Mars that, once again, looked altogether different from anything that had been drawn before. With his map, though, things settled down. Finally, he saw what there really was to see, or so it seemed; for later astronomers over the next 100 years saw generally what he had seen in the way of a pattern of light and dark areas.

By that time, though, Maxwell and Boltzmann had come out with their kinetic theory of gases, and it didn't seem that a body with the mass and gravitational field of Mars ought to have large, open bodies of water. Even at Mars's low temperature, water vapour must have found it too easy to escape, if the atmosphere were thinner than Earth's. The suspicion grew, therefore, that Mars must be water poor. It had its ice caps, to be sure, and it might have its marshy and boggy regions – but open seas and oceans seemed unlikely.

What, then, were the dark areas?

They might be areas of vegetation, growing in the boggy regions, while the light areas were sandy desert. It was interesting that when it was summer in a particular hemisphere, and the ice cap shrank as it presumably melted, the darkened areas became more extensive as though the melting ice irrigated the soil and allowed vegetation to spread.

Many people began to take it for granted that Mars was the abode of life.

In the course of his observations of Mars in 1877, moreover, Schiaparelli noticed there were rather thin dark lines present on Mars, each of which connected two larger dark areas. These had been noticed back in 1869 by another Italian astronomer, Pietro Angelo Secchi (1818–1878). Secchi had called them channels, a natural name for a long thin body of water connecting two larger bodies. Schiaparelli used the same term. Both Secchi and Schiaparelli naturally used the Italian word for channels, which is *canali*.

Schiaparelli's *canali* were longer and thinner than those Secchi had reported seeing, and they were more numerous. Schiaparelli saw about forty of them and included them on his map, giving them the names of rivers in ancient history and mythology.

Schiaparelli's map and his *canali* were greeted with great interest and enthusiasm. Nobody besides Schiaparelli had seen the *canali* in the course of the 1877 observations, but afterwards astronomers started looking for them in particular and some reported seeing them.

What's more, the word *canali* was translated into the English word *canals*. That was important. A channel is any narrow waterway, and is usually a naturally formed body of water. A canal, however, is a narrow, artificial waterway constructed (on Earth) by human beings. As soon as Englishmen and Americans began calling the *canali* canals instead of channels, they began automatically to think of them as being artificial and therefore as having been built by intelligent beings.

At once there came to be enormous new interest in Mars. It was the first time (so it seemed) that scientific evidence had been advanced that strongly favoured the existence of extra-terrestrial intelligence.

The picture created was of a planet that was older than Earth and that was slowly losing its water because of the weakness of its gravitational field. The intelligent Martians, with a longer history than ours and with a more advanced technology, faced death by desiccation.

Heroically, they strove to keep the planet alive. They built huge canals to transport needed water from the last planetary reservoir, the ice caps. It was a very dramatic picture of an ancient race of beings, perhaps a dying species, who refused to give up and who kept their world alive by resolution and hard work. For nearly a century, this view remained popular with many people, and even with a few astronomers.

There were astronomers who added to Schiaparelli's reports. The American astronomer William Henry Pickering (1858–1938) reported round dark spots where canals crossed, and these were called oases. Flammarion, who was a great believer in extraterrestrial life, as I said before, was particularly enthusiastic about the canals. In 1892, he published a large book called *The Planet Mars*, in which he argued in favour of a canal-building civilization.

By far the most influential astronomer who supported the notion of a Martian civilization was the American Percival Lowell (1855–1916). He was a member of an aristocratic Boston family and he used his wealth to build a private observatory in Arizona, where the mile-high dry desert air and the remoteness from city lights made the visibility excellent. The Lowell Observatory was opened in 1894.

For fifteen years, Lowell avidly studied Mars, taking thousands of photographs of it. He saw many more canals than Schiaparelli ever did, and he drew detailed pictures that eventually included over five hundred canals. He plotted the oases at which they met, recorded the fashion in which the individual lines of particular canals seemed to double at times, and studied the sensational changes of light and dark that seemed to mark the ebb and flow of agriculture. He was completely convinced of the existence of an advanced civilization on Mars.

Nor was Lowell bothered by the fact that other astronomers couldn't see the canals as well as he. Lowell pointed out that no one had better seeing conditions than he had in Arizona, that his telescope was an excellent one, and that his eyes were equally excellent.

In 1894, he published his first book on the subject, *Mars*. It was well written, clear enough for the general public, and it supported the notion of an ancient, slowly dying Mars; of a race of advanced engineers keeping the planet alive with gigantic irrigation projects; of canals marked out and made visible from Earth by the bands of vegetation on both borders.

Lowell's views were even more extreme in later books he published – *Mars and Its Canals* in 1906 and *Mars as the Abode of Life* in 1908. The general public found the whole thing exciting, for the thought of a nearby planet populated by an intelligence advanced beyond that of human beings was dramatic.

Lowell's role in making advanced Martian life popular was outpaced, however, by the English fiction writer, H. G. Wells.

In 1897, Wells published a novel, *War of the Worlds*, in serial form in a magazine, and the next year it appeared in book form. It combined the view of Mars as presented by Lowell with the situation as it had existed on Earth over the preceding twenty years.

In those decades, the European powers – chiefly Great Britain and France, but including also Spain, Portugal, Germany, Italy, and Belgium – had been carving up Africa. Each nation established colonies with virtually no regard for the wishes of the people already living there. Since the Africans were dark skinned and had cultures that were not European, the Europeans considered them inferior, primitive, and barbarous, and felt they had no rights to their own territory.

It occurred to Wells that if the Martians were as far advanced scientifically over Europeans as Europeans were over Africans, the Martians might well treat Europeans as Europeans treated Africans. *War of the Worlds* was the first tale of interplanetary warfare involving Earth.

Until then, tales of visitors to Earth from outer space had pictured those visitors as peaceful observers. In Wells's novel, however, the outsiders came with weapons. Fleeing a Mars on which they could barely keep alive, they arrived at lush, watery Earth and prepared to take over the planet to make a new home for themselves. Earth people were merely animals to them, creatures whom they could destroy and devour. Nor could human beings defeat the Martians or even seriously interfere with them, any more than the Africans could deal with the European armed forces. Though the Martians were defeated in the end, it was not by human beings, but by Earthly decay bacteria, which the Martians' bodies were not equipped to resist.

It proved a popular novel and set off a wave of imitations, so that for the next half-century human beings took it for granted that any invasion of extraterrestrial intelligence would lead to the extermination of humanity.

On 30 October 1938, for instance, nearly forty years after *War of the Worlds* was published, Orson Welles (born 1915), only twenty-three years old at the time, produced a radio dramatization of the story. He chose to bring the story up to date, and had the Martians land in New Jersey rather than in Great Britain. He told the events in as realistic a fashion as possible, with authentic-sounding news bulletins, eyewitness reports, and so on.

Anyone who had turned the programme on at the start would have been informed that it was fiction, but some weren't listening closely enough and others turned it on after the start and were transfixed at the events that were apparently taking place – especially those who were near the sites of the reported landings.

A surprising number of people did not pause to question whether it was at all likely that there was an invasion of Martians, or whether there were even Martians at all. They took it for granted that Martians existed and had arrived to conquer Earth and were succeeding. Hundreds got into their automobiles and fled in terror. Like the Moon Hoax of just a century before, it was a remarkable example of how ready people were to accept the notion of extraterrestrial intelligence.

Though Lowell and his theories concerning the Martian canals were successful with the general public, professional astronomers were extremely doubtful. At least the large majority were.

A number insisted that though they looked at Mars carefully, they never saw any canals, and they were not soothed by Lowell's lofty assurance that their eyes and telescopes just weren't good enough. The American astronomer Asaph Hall (1829–1907), whose eyes had been good enough in 1877 to discover the tiny Martian satellites, never saw a canal.

One American astronomer, Edward Emerson Barnard (1857–1923), was a particularly keen observer. In fact, he is often cited as the astronomer with the sharpest eyes on record. In 1892, he discovered a small fifth satellite of Jupiter, one that was so small, and so close to the brightness of Jupiter itself, that to see it required eyes of almost superhuman keenness; yet Barnard insisted that no matter how carefully he observed Mars, he could never see any canals. He said flatly that he thought it was all an optical illusion; that small, irregular patches of darkness were made into straight lines by eyes straining to see objects at the very edge of vision.

This notion was taken up by others. An English astronomer, Edward Walter Maunder (1851–1928), even put it to the test in 1913. He set up circles within which he put smudgy irregular spots and then placed schoolchildren at distances from which they could just barely see what was inside the circles. He asked them to draw what they saw, and they drew straight lines such as those Schiaparelli had drawn of the Martian canals.

Meanwhile, astronomers continued to study the habitability of Mars. As the twentieth century advanced, instruments were devised that could detect and measure tiny quantities of heat. If these heat detectors were placed at the focus of a telescope and the light from Mars were allowed to fall upon it, the temperature of Mars could be deduced.

This was first done in 1926 by two American astronomers, William Weber Coblentz (1874–1962) and Carl Otto Lampland (1873–1951). From such measurements, it seemed that at the Martian equator the temperature would rise above the melting point of ice at times. In fact, it was even possible for the equatorial temperatures to rise as high as 25°C (77°F) on rare occasions.

The temperature dropped sharply during the night, however. There was no way of following the temperature at night, for the night side of Mars was always on the side away from Earth. However, the temperature of the early morning could be taken at the western edge of the Martian globe where the surface of the planet was just emerging from night and into the dawn. After twelve and a quarter hours of dark, the temperature could be as low as –100°C (–150°F).

In short, it looked as though the temperature of Mars was too low for water to exist as anything but ice, except in a narrow region around the equator and for brief times around midday. Elsewhere, the climate on Mars was colder than that in Antarctica.

Worse yet, the great difference between dawn temperatures and noon temperatures meant that the Martian atmosphere was probably thinner than had been thought till then. An atmosphere acts as a blanket, absorbing and transferring heat, and the thinner it is the more rapidly temperatures go up and down.

What's worse is that a thin atmosphere does not absorb much of the energetic radiation of the Sun. On Earth, the relatively thick atmosphere acts as an efficient blanket absorbing the energetic radiation that bombards our planet from the Sun and elsewhere.

All these energetic radiations would be fatal to unprotected life if they fell upon Earth's surface in full strength. Mars is farther from the Sun than we are and it receives a smaller concentration of ultraviolet light, for instance. However, that smaller concentration reaches the Martian surface in far greater quantities, it would appear, than it reaches the terrestrial surface.

By the 1940s, it became possible to analyse the infrared radiation from Mars to analyse the content of its atmosphere. This was done in 1947 by the Dutch-American astronomer Gerard Peter Kuiper (1905–1973). He found that what little there was of the Martian atmosphere was almost entirely carbon dioxide. There was very little water vapour and apparently no oxygen at all.

Considering the frigidity of Mars, some astronomers began to wonder if there was any water on Mars at all. Might the ice caps not be frozen water, but frozen carbon dioxide instead?

Taking all things into consideration – a thin atmosphere of carbon dioxide, ultraviolet light bombarding the Martian surface, temperatures of deep frigidity – it seemed unlikely that the kind of complex life forms one would expect to have developed intelligence would have evolved on Mars.

The feeling grew that if the canals existed at all, they were natural phenomena, not the product of a race of advanced engineers.

But then, if not intelligent life, what about primitive life? On Earth, there are bacteria that can live on chemicals poisonous to other forms of life. There are lichens that can grow on bare rock, and on mountaintops where the air is so thin and the temperature is so low that one might almost imagine one's self to be on Mars.

Beginning in 1957, experiments were conducted to see if any simple life forms that were adapted to severe conditions on Earth might survive in an environment that, as far as possible, duplicated what was then known of the Martian environment.

Over and over again it was shown that some life forms would survive.

Perhaps, in that case, we ought not abandon all hope of complex life forms either. After all, life on Earth has evolved to fit the terrestrial environment. To us, therefore, conditions on Earth seem pleasant, and conditions that are considerably different from those on Earth seem unpleasant. On Mars, how-ever, life forms would have evolved to suit the conditions there, and it would then be those conditions that would seem pleasant to them.

The question appeared moot right into the 1960s.

Mars probes

In the 1960s, rocket-powered probes were being launched that were intended to pass near the planet and send back informa-tion (like the ones I have already mentioned in connection with Mercury and Venus).

On 28 November 1964, the first successful Mars probe, *Mariner 4*, was launched. As *Mariner 4* passed Mars it took a series of twenty photographs that were turned into radio sig-nals beamed back to Earth, where they were turned into photographs again.

What did they show? Canals? Any signs of a high civiliza-tion or, at least, of life?

What the photographs showed turned out to be completely unexpected, for as they were received, astronomers saw what were clearly craters – craters that looked very much like those on the Moon.

The craters, at least as they showed up on the *Mariner 4* pictures, seemed so many and so sharp that the natural con-clusion was that there had been very little erosion. That seemed to mean not only thin air, but very little life activity. The craters shown in the photographs of *Mariner 4* seemed to be the mark of a dead world.

Mariner 4 was designed to pass behind Mars (as viewed

from Earth) after its flyby, so that its radio signals would eventually pass through the Martian atmosphere on their way to Earth. From the changes in the signals, astronomers could deduce the density of the Martian atmosphere.

It turned out that the Martian atmosphere was even thinner that the lowest estimates. It was less than $\frac{1}{100}$ as dense as Earth's atmosphere. The air pressure at the surface of Mars is about equal to that of Earth's atmosphere at a height of 32 kilometres (19 miles) above the Earth's surface. This was another blow to the possibility of advanced life on Mars.

In 1969, two more rocket probes, *Mariner 6* and *Mariner 7*, were sent past Mars. They had better cameras and instruments, and took more photographs. The new and much better photographs showed that there was no mistake about the craters. The Martian surface was riddled with them – as thickly, in places, as the Moon.

The new probes, however, showed that Mars was not entirely like the Moon. There were regions in the photographs in which the Martian surface seemed flat and featureless and others where the surface seemed jumbled and broken in a way that was not characteristic of either Moon or Earth. There were still no signs of canals.

On 30 May 1971, *Mariner 9* was launched and sent on its way to Mars. This probe was not merely going to pass by Mars, it was to go into orbit about it. On 13 November 1971, it went into orbit. Mars was at that time in the midst of a planet-wide dust storm and nothing could be seen, but *Mariner 9* waited. In December 1971, the dust storm finally settled down and *Mariner 9* got to work taking photographs of Mars. The entire planet was mapped in detail.

The first thing that was settled, once and for all, was that there were no canals on Mars. Lowell was wrong after all. What he had seen was an optical illusion.

Nor were the dark areas either water or vegetation. Mars seemed all desert, but here and there one found dark streaks that usually started from some small crater or other elevation. They seemed to be composed of dust particles blown by the wind and tended to collect where an elevation broke the force of the wind, on the side of the elevation away from the wind.

There were occasional light streaks, too, the difference between the two resting perhaps in the size of the particles. The possibility that the dark and light areas were differences in dust markings and that the dark areas expanded in the spring because of seasonal wind changes had been suggested a few years earlier by the American astronomer Carl Sagan (born 1935). *Mariner 9* proved him to be completely correct.

Only one of the hemispheres of Mars was cratered and Moonlike; the other was marked by giant volcanoes and giant canyons, and seemed geologically alive.

One feature of the Martian surface roused considerable curiosity. These were markings that wiggled their way across the Martian surface like rivers and that had branches that looked for all the world like tributaries. Then, too, both polar ice caps seemed to exist in layers. At the edge, where they are melting, they looked just like a slanted stack of thin poker chips.

It is possible to suppose that Mars's history is one of weather cycles. It may now be in a frigid cycle, with most of the water frozen in the ice caps and in the soil. In the past, and possibly again in the future, it may be in a mild cycle, in which the ice caps melt, releasing both water and carbon dioxide, so that the atmosphere becomes thicker and the rivers grow full.

In that case, even if there is no apparent life on Mars now there may have been in the past, and there may again be in the future. As for the present, life forms could be hibernating in the frozen soil, in the form of spores.

In 1975, two probes, *Viking 1* and *Viking 2*, the former launched on 20 August, the latter on 9 September, were sent to Mars. They were to land on the planet and observe it in various ways. In particular, they were to test the planet for signs of life.

They landed safely in the summer of 1976 in two widely separated places. They analysed the Martian soil and found it to be not too different from Earth's, but richer in iron and less rich in aluminium.

Three experiments were conducted that might detect life. All three gave results of the kind that might be expected if there were living cells in the soil.

There was, however, a fourth experiment that cast doubt upon the first three. To understand that, we will have to consider the nature of the molecules most characteristic of living organisms as we know it.

Against the background of water, there is in living organisms a rapid and never ending interplay involving complex molecules made up of anywhere from a dozen to a million atoms. These are found, in nature, only in living organisms and in the dead remnants of once living organisms.* For that reason, such complex molecules are called organic compounds.

Organic compounds have something in common – the element carbon. Carbon atoms have a unique facility for combining with each other in complex chains, both straight and branched, and in rings or collections of rings to which chains of atoms, either straight or branched, can be attached. Also attached on the outskirts of the carbon chains and rings are atoms and combinations of atoms of other elements, chiefly those of hydrogen, oxygen, and nitrogen, plus occasional atoms of sulphur, phosphorus, and so on. Sometimes, one of these other atoms may actually be incorporated into the body of the carbon chain or ring.

No type of atom other than carbon can form chains and rings with anything like this facility.

Furthermore, it is difficult to imagine that so complex and versatile a phenomenon as life can make do with anything less complex than the molecules with which we are familiar in Earthly organisms.

This does not seriously limit the infinite variability of life. It

* They can also be formed in the laboratory. In addition, uncounted thousands of such compounds, not quite like any to be found in living organisms or their residues, have also been synthesized by chemists. But then, chemists are living organisms so that even the synthetic molecules that 'are not found in nature' are the result of the actions of living organisms.

is enormously variable here on Earth, in form, in structure, in behaviour, in adaptation, yet it is all based on organic compounds, which are in turn based on chains and rings of carbon atoms.

What is more, the number of conceivable variations on the structure of the organic compounds is so enormous as to be far beyond expression in any comprehensible manner. The number of organic compounds used by terrestrial life compared to all the organic compounds there can conceivably be is *far less* than the size of an atom compared to the size of the entire Universe.

In summary, then, the number of complex compounds based on carbon atoms is virtually limitless, and in comparison the number of complex compounds that do not contain the carbon atom is virtually zero. We can assume, therefore, that if a world lacks organic compounds, it lacks life.

Again, it would be well not to hasten on too rapidly. Can we be sure that under certain conditions of a type with which we are not familiar, elements or combinations of elements other than carbon might not produce complicated compounds? Can we be sure that under certain conditions life might not be built up out of relatively simple compounds?

We can't be. Considering how little we know of the details of other worlds, and of the finer points of life other than what we can glean from our own example, we can't be sure of anything.

But we *can* ask for evidence. There is no evidence whatever of the possible existence of molecules as complex, delicate, and versatile as organic compounds, built up of any element but carbon, or of any combination of elements that excludes carbon. Nor is there any evidence that something as complex as life could be built up out of relatively simple compounds.

Therefore, *until evidence to the contrary is forthcoming*, we can only assume that if organic compounds are not present, life is not present.

As it happens, the analysis of Martian soil by *Vikings 1* and *2* indicates the absence of organic compounds.

This leaves the matter of life on Mars ambiguous. The evidence is clear-cut neither for nor against and must await further and better testing. Nevertheless, if life is present, there seems very little chance that it is anything more than very primitive in nature – no more than on the level of bacterial life on Earth.

Such simple life would be quite sufficient to excite biologists and astronomers, but as far as the search for extraterrestrial intelligence is concerned, we are left with what is overwhelmingly likely to be zero.

We must look elsewhere.

4 The outer Solar System

Planetary chemistry

The inner Solar System out to the orbit of Mars is a comparatively small structure. Beyond Mars is the 'outer Solar Systew', which is far vaster and within which giant planets orbit. There are no less than four such giants out there: Jupiter, Saturn, Uranus, and Neptune. Each dwarfs Earth, particularly Jupiter, which has over 1,000 times the volume of Earth and over 300 times its mass.

Why should the inner Solar System contain pygmies and the outer Solar System giants? Consider:

The cloud out of which the Solar System was formed would naturally have been made up of the same kind of substances that make up the Universe generally – more or less. Astronomers have, through spectroscopy, determined the chemical structure of the Sun and of other stars, as well as of the dust and gas between the stars. They have therefore come to some conclusions as to the general elementary makeup of the Universe. This is given in the table on p. 76. As you see, the Universe is essentially hydrogen and helium, the two elements with the simplest atoms. Together hydrogen and helium make up nearly 99.9 per cent of all the atoms in the Universe. Hydrogen and helium are, of course, very light atoms, not nearly as heavy as the others, but they still make up about 98 per cent of all the mass in the Universe.

The fourteen most common elements given in the table overleaf make up almost the entire Universe. Only one atom out of a quarter million is anything else.

Of the fourteen, the atoms of helium, neon, and argon do not combine either with each other or with the atoms of other elements.

Element	Number of Atoms for every 10,000,000 Atoms of Hydrogen
Hydrogen	10,000,000
Helium	1,400,000
Oxygen	6,800
Carbon	3,000
Neon	2,800
Nitrogen	910
Magnesium	290
Silicon	250
Sulphur	95
Iron	80
Argon	42
Aluminium	19
Sodium	17
Calcium	17
all other elements combined	50

Hydrogen atoms will combine with other atoms after colliding with them. In view of the makeup of the Universe, however, hydrogen atoms will, if they collide with anything at all, collide with other hydrogen atoms. The result is the formation of hydrogen molecules, made up of two hydrogen atoms each.

Oxygen, nitrogen, carbon, and sulphur are made up of atoms that are likely to combine with hydrogen atoms when the latter are present in overwhelming quantity. Each oxygen atom combines with two hydrogen atoms to form molecules of water. Each nitrogen atom combines with three hydrogen atoms to form molecules of ammonia. Each carbon atom combines with four hydrogen atoms to form molecules of methane. Each sulphur atom combines with two hydrogen atoms to form hydrogen sulphide.

These eight substances – hydrogen, helium, neon, argon, water, ammonia, methane, and hydrogen sulphide – are all gases at Earth temperatures or, in the case of water, an easily vaporized liquid. We can lump them all together as 'volatiles' (from a Latin word for *to fly* since, as gases or vapours, they are not held firmly to matter, but tend to diffuse or fly away).

Silicon combines with oxygen much more easily than with hydrogen. Magnesium, aluminium, sodium, and calcium combine readily with the silicon-oxygen combination, and these six elements together make up the lion's share of the rocky materials ('silicates') that we are familiar with.

As for iron – that tends to be present in rocks, but is sometimes present in considerable excess so that much of it remains in metallic form. To the iron are added the similar but less common metals nickel and cobalt.

The atoms and molecules of rocks and metals cling together, bound by strong chemical forces, so that they remain solid up to white-hot temperatures. They do not require gravitational forces to hold them together so that atoms in tiny grains of rock or metal, where the gravitational forces are utterly negligible, nevertheless hold firmly together.

Of the original material composing the primordial nebula out of which the Solar System was formed about 99.8 per cent of the mass were volatiles, and only 0.2 per cent were solids.

In the inner Solar System, the heat of the nearby Sun raised the temperature high enough to keep the atoms and molecules of the volatiles moving fast enough to be too nimble to be caught gravitationally. The planets in the inner Solar System ended up composed of rocks and metals that required no gravitational force to be held, but that also made up only a very small part of the nebular material. That is why the inner planets are small.

The smallest, in fact, contain no volatiles at all. Mercury is made up of a sizeable metal core, surrounded by a rocky mantle. (We know this is so because Mercury's density is so high that much of it must be the high-density metal and only the rest of it medium-density rock.) The Moon is made up of rock only. Its density is too small to allow any metal core of significant size. Both Mercury and the Moon lack volatiles.

Mars, like the Moon, is of rock only. Earth and Venus, like Mercury, are made up of rock over a metal core. These three, however, are all large enough to be able to retain some volatiles by gravitational attraction.

Beyond the orbit of Mars it becomes easier to accumulate volatiles at a given level of gravitational intensity. For one thing, at lower temperatures, all molecules move more slowly and are less likely to exceed escape velocity. For another, the volatiles solidify one by one as the temperature drops, and solid volatiles will cling together by chemical attraction and no longer be dependent on gravitational pull.

The freezing points, under terrestrial conditions, of the eight volatiles are given in the accompanying table:

Substance	Freezing point		
	° Celsius	° Fahrenheit	° Absolute
Water	0.0	32.0	273.1
Ammonia	−77.7	−82.3	195.4
Hydrogen sulphide	−85.5	−96.3	187.6
Methane	−182.5	−270.11	90.6
Argon	−189.2	−283.0	83.11
Neon	−248.7	−390.1	24.4
Hydrogen	−259.1	−408.8	14.0
Helium (under pressure)	−272.2	−432.4	0.11

This means that anywhere beyond the orbit of Mars even small bodies can collect not only metal and rock, but also such volatiles as water, ammonia, and hydrogen sulphide in solid form. If the small bodies are sufficiently far from the Sun to have temperatures very low, then methane and argon can also be collected in solid form. Neon, hydrogen, and helium freeze at so low a temperature that a small body, even right out at the known limits of the Solar System, cannot collect them.

Frozen water is, of course, ice. The solid forms of the other volatiles resemble ice in physical appearance so that the solid volatiles may be referred to as ice. To distinguish the original ice, frozen water, we may call it water-ice.

Titan

Let us see, then, how little we can know about a world in the outer Solar System, and still be able to judge at once that it cannot bear life (as we know it).

We have already decided that organic compounds are essential for life. Organic compounds consist of molecules made up of chains and rings or carbon atoms to which are invariably added hydrogen atoms, with lesser admixtures of nitrogen atoms, oxygen atoms, and sulphur atoms. These five types of atoms make up ninety-nine per cent or more of all the atoms in organic compounds. These atoms also make up five of the eight volatile substances. (The atoms of the other three – argon, neon, and helium – undergo no combinations and play no role in life.)

It is clear, then, that life as we know it is a function of the volatiles and that no world can bear life unless it has at least some volatile matter.

At the temperatures prevailing beyond the orbit of Mars, almost any body, however small, can contain some volatile matter. Every once in a while, for instance, a meteorite falls that is found to contain water, hydrocarbons,* and other volatiles. Not much, only up to five per cent or so – but they're there.

Such meteorites, called carbonaceous chondrites, are few indeed compared to the ordinary meteorites that are constructed of metal, or of rock, or of a mixture of the two. Indeed, only about twenty carbonaceous chondrites have ever been located.

This does not really mean that carbonaceous chondrites are rare. They could be very common. However, they tend to be structurally weaker than the rocky and metallic meteorites. The carbonaceous chondrites crumble away more easily in the

* Substances with molecules made up of carbon and hydrogen atoms only. Methane is an example.

white-hot passage through the atmosphere, so that very few fragments of any of them survive to strike Earth's surface.

In recent years, it has turned out that most of the asteroids, particularly those farther from the Sun, have the characteristics (dark colour and low density) of the carbonaceous chondrites and therefore have volatile material in them. The two small satellites of Mars are much darker than Mars itself in colour and are lower in density, so they must contain some volatile matter.

Then, too, there are the comets, which exist as small solid bodies in that part of their orbit far from the Sun. They are perhaps only a few kilometres in diameter and are largely or almost entirely composed of icy materials.

When they pass through the part of the orbit in the neighbourhood of the Sun, some of the ices vaporize and liberate granules of rock or metal that may be mixed with the ices. The whole forms a misty 'coma' about the still solid 'nucleus'. The Sun constantly emits streams of rapid subatomic particles in all directions (the 'Solar wind') and this sweeps the coma outwards in a direction away from the Sun, forming a long, wispy 'tail'.

Any objects in the outer Solar System that are larger than asteroids and comets would contain volatile matter almost as a matter of course, we might reason.

Although a lack of volatile-materials is a sure sign that the world does not contain life (as we know it), the converse is not true. A world may possess volatile materials and yet not contain life (Venus is an example). If this were not so, we would have to judge that just about every object beyond Mars was life bearing.

After all, volatile materials might be present, yet organic compounds of sufficient complexity to make life possible might not form.

From our vantage point on Earth, however, it is not easy to tell whether a small body beyond the orbit of Mars contains complex organic compounds or not. Short of exacting detail beyond our capacities to do so, is there any way of judging whether life is likely to be present or absent in a distant world?

We can begin by pointing out that we have already said that a liquid medium, like that of water, is required for life.

If, however, a world has sufficient liquid on its surface to make possible the presence of life – not merely as a thin scattering of bacterialike organisms, but in sufficient complexity to allow an approach to intelligence – this liquid would surely vaporize to some extent.

If the world was not capable of holding on to the vapour through its gravitational force, then the liquid would continue vaporizing until it was all gone. If the world *were* capable of holding on to the vapour, then it would have an atmosphere of more than traces of gas; an atmosphere consisting of that vapour at the very least, and possibly of other gases as well.

It follows, then, that a world without an atmosphere cannot bear life (as we know it) above the bacterial level; not because the atmosphere is itself necessarily essential to life, but because sizeable quantities of free liquid on the surface are necessary for more-than-bacterial life. Without an atmosphere, what volatiles are present must be in the frozen, solid state, and that is insufficient for life.

With this in mind, let's consider those objects that lie beyond the orbit of Mars and that are less than 2,900 kilometres (1,800 miles) in diameter.

There are uncounted numbers of these, trillions upon trillions of dust grains, billions of comets, tens of thousands of asteroids, and a couple of dozen small satellites. All can be eliminated. Although a very large proportion of them, perhaps almost all of them over the size of dust grains, contain volatile material, none has a permanent atmosphere or any hope of free liquid. Those comets that approach the Sun have a temporary atmosphere during the approach, but it is very doubtful that they have free liquid even then – and the period of atmosphere makes up a very small fraction of their total stay in orbit.

What about the objects beyond the orbit of Mars that have

diameters between 2,900 and 6,500 kilometres (1,800 and 4,000 miles)?

There are exactly six of these, the satellites, Io, Europa, Ganymede, Callisto, Titan, and Triton. (Until 1978 it was thought the planet Pluto was a seventh, but very recent information makes it appear a surprisingly small body.)

Of these six bodies, the four satellites Io, Europa, Ganymede, and Callisto circle Jupiter and are the nearest to the Sun. None has anything better than trace atmospheres.

Io, which is the closest to Jupiter, must have been exposed to considerable warmth in the early days of planetary formation when Jupiter itself, as it formed, radiated heat strongly. At any rate, judging from its density Io is very much like our Moon and includes little if any volatile material in its structure.

The farther satellites have progressively lower densities and must, therefore, contain more and more volatiles. These volatiles must be chiefly water, together with smaller quantities of ammonia and hydrogen sulphide. Methane is a gas even at temperatures as low as those that prevail in the neighbourhood of Jupiter, and its molecules are too nimble to be held by the small gravitational pulls of the satellites.

Europa, the second of the large satellites, probably has a thin layer of water-ice on its surface. The third and fourth of the large satellites, Ganymede and Callisto, have much thicker layers of volatile materials around a rocky core. The layers may even be hundreds of kilometres thick. On the surface, there is a layer of water-ice but underneath, warmed by internal heat, there may be a layer of liquid water. Can life have developed on these two satellites in a region of eternal darkness, sealed away from the rest of the Universe by an unbroken miles-thick layer of ice? As yet, we can't say.

If Jupiter's satellites are the nearest of the six bodies we are discussing, Pluto lies beyond all six. Pluto is so far from the Sun and is at such a low temperature that even methane is frozen. Recent observations of the light it reflects indicate, in fact, that it is covered with a layer of frozen methane. It might conceivably have a thin atmosphere of hydrogen, helium, and neon, but there is as yet no indication of that. Even if it did, however, this would not help it have any free liquid on its surface, since at Pluto's temperature, hydrogen, neon, and helium are gases and everything else is solid. Furthermore, in 1978 it was found that Pluto was not one body, but two. It has a satellite, now named Charon, and each – the planet and the satellite – is smaller than our Moon. Neither can bear life.

The next-farthest world is Triton, a satellite of the planet Neptune. Very likely it is in Pluto's case, with a coating of solid methane and a very thin atmosphere of hydrogen, neon, and helium, but as yet that is only a presumption.

The remaining world in this size range is Titan, the largest satellite of Saturn. It is farther from the Sun and colder than the four satellites of Jupiter. It is closer to the Sun and warmer than Triton, Charon, and Pluto.

Titan's temperature is about –150°C (–207°F), 15 Centigrade degrees lower than that of Jupiter's satellites. At Titan's temperature methane is still gaseous, but it is pretty close to the point where it would liquify (–161.5°C or –233.1°F) and its molecules are sluggish indeed. They could be held by Titan's gravitational pull, even though that pull is only two-thirds as intense as that of our Moon.

It follows that Titan could conceivably have a methane atmosphere and, in 1944, Gerard Kuiper actually detected such an atmosphere. What is more, the atmosphere is a substantial one, very likely denser than that of Mars.

Titan is the only satellite in the Solar System known to have a true atmosphere. It is also the smallest body in the Solar System to have a true atmosphere, and it is the only body of any size to have an atmosphere that is primarily methane.

Methane, with a molecule consisting of one carbon atom and four hydrogen atoms, is the smallest organic compound. Thanks to the peculiar properties of the carbon atom and the readiness with which it will hook on to other carbon atoms, it is easy for methane molecules to combine into larger ones containing two carbon atoms, or three or four, with some appropriate number of hydrogen atoms also attached. The Sun, although very distant from Titan, would nevertheless supply enough energy to drive such reactions.

It may, therefore, turn out that Titan's atmosphere has as minor constituents a complicated mix of vapours of higher hydrocarbons and it may be this mix that causes Titan to appear distinctly orange in colour when viewed through the telescope.

The more complicated a hydrocarbon molecule, the higher the temperature at which it liquefies. Though the higher hydrocarbons may exist as vapours in the atmosphere, the major portion will be in liquid form on the surface. Since cigarette lighter fluid is made up of molecules of hydrocarbon with five or six carbon atoms, we might visualize Titan as possessing lakes and seas of cigarette lighter fluid, with still more complicated molecules dissolved in them, or forming sludges along the shores of those lakes and seas.

Thus, Titan would have free liquid in quantity *and* organic compounds in quantity as well.

This represents the minimum requirement for life, but there is a serious question as to whether hydrocarbons can substitute for water as the basic liquid against which the pattern of life can be constructed.

Water is a 'polar liquid'. That is, its molecules are asymmetric and there are tiny electric charges at the opposite ends. These tiny electric charges set up attractions and repulsions that play an important part in the chemical changes characteristic of life. Hydrocarbon molecules are 'nonpolar liquids', however, with symmetrical molecules and no tiny electric charges. Can nonpolar liquids serve as an adequate background for life?

Can *any* liquid other than water serve as a background to life? The only liquids that have any reasonable chance to do so are those that are present in large quantities in the Universe generally and that are indeed liquid at planetary temperatures. In addition to water and hydrocarbons there are only two other candidates, ammonia and hydrogen sulphide. Ammonia is a polar liquid, but not as polar as water, and hydrogen sulphide is less polar still.

With sufficient ingenuity we can work out chemistries that use these liquids as background and have life in the foreground, but those are just exercises in speculation. We have no evidence whatsoever that any common liquid will substitute for water.

Until such evidence is forthcoming, at least some tiny scrap of it, we must remain conservative and count on water life only. For that reason, although Titan will offer us a fascinating chemical world if we can ever study it in some detail, we cannot bet very heavily on it as an abode of life.

Jupiter

In the cold reaches beyond Mars, it might happen that a world as it formed would pick up enough in the way of icy materials (in addition to what rock and metal might be available) to develop a gravitational field strong enough to hold on to helium and neon. The added mass would intensify the gravitational field and make it possible, perhaps, for it to hold on to hydrogen, which is present in greater amounts than any other substance.

Every bit of hydrogen added makes it that much easier to gather more hydrogen, so that there is a snowball effect that quickly empties surrounding space of its material and produces a giant planet, leaving only enough material behind to make small bodies such as satellites and asteroids.

There are four planets in the outer Solar System that have been formed in this way: Jupiter, Saturn, Uranus, and Neptune.

Of these, the largest is Jupiter, with a diameter of 143,200 kilometres (89,000 miles) or 11.23 times that of Earth. The smallest is Neptune, with a diameter of 49,500 kilometres (30,800 miles) or 3.88 times that of Earth. The volumes range from 1,415 times that of Earth for Jupiter to 58 times that of Earth for Neptune.

Because these outer giants are made up so largely of the volatiles, which are of low density, their overall density is considerably smaller than that of Earth. The densest of the giants is Neptune, which has an average density of 1.67 times that of water. The least dense is Saturn, with an average density 0.71 times that of water. (Saturn would float on water if there were an ocean big enough and if Saturn would remain intact in the process.) Compare this with Earth's average density of 5.5 times that of water.

Since the outer giants are so low in density, their mass (the quantity of matter they contain, roughly speaking) is lower than one might think from their size. The most massive is Jupiter, with 318 times the Earth's mass; and the least massive is Uranus, with 14.5 times the Earth's mass.

From such considerations alone, it is clear that the properties and nature of the outer giants is enormously different from Earth's. Is life conceivable on them?

On 2 March 1972, a probe, *Pioneer 10*, was launched for a rendezvous with Jupiter. On 3 December 1973, it passed Jupiter at a distance of only 135,000 kilometres (85,000 miles) from its surface.

During the four days it took *Pioneer 10* to fly by Jupiter, its instruments picked up radiation, counted particles, measured magnetic fields, noted temperatures, and analysed sunlight passing through Jupiter's atmosphere.

After *Pioneer 10* had triumphantly passed Jupiter, a second probe, *Pioneer 11*, a close duplicate of the first one, was approaching the planet. It had left Earth on 5 April 1973, and passed Jupiter at a distance of 42,000 kilometres (26,000 miles) from its surface on 2 December 1974. It passed over Jupiter's north polar region, which human beings cannot see from Earth.

Both probes sent back photographs and other useful information. From that information, astronomers feel that rock and metal make up a very small quantity of Jupiter's total structure. Apparently, Jupiter would seem to consist chiefly of hydrogen, with a small admixture of helium, and traces (in comparison) of the other volatiles. Just as Earth is essentially a spinning ball of rock and metal, so Jupiter is a spinning ball of hot liquid hydrogen. (Ordinarily, liquid hydrogen boils at extremely low temperatures, but under the enormous pressures within Jupiter it apparently reaches far higher temperatures.)

The outermost skin of Jupiter's ball of liquid is cold, but the temperature rises rapidly with depth. At 950 kilometres (600 miles) below the visible cloud surface, the temperature is already 3,600°C (6,500°F).

In the uppermost cool layer of the planet there is water, ammonia, methane, and other volatiles, including small percentages of hydrocarbons with two or three carbon atoms in the molecule.

Naturally, there is probably circulation in the planetary liquid of Jupiter as there is in Earth's oceans. There may be vast columns of the Jupiter-liquid sinking and warming, while other columns, equally vast, are rising and cooling.

Here the arguments for life are intriguing. Water is certainly present in the fluid, and while it may be present in small percentages, on vast Jupiter even a small percentage is a large quantity in absolute terms. Even though the water is completely overwhelmed by the hydrogen, there could be more water by far on Jupiter than on Earth.

Then, too, there is methane and ammonia in addition to water, and the three could combine to form the kind of organic molecules we associate with life. It would take energy to force the combination, but considering Jupiter's enormous internal heat, that would be no problem.

We could easily imagine living cells, and perhaps complicated multicellular animals, living in the Jovian ocean, maintaining themselves at a level of comfortable temperature, swimming up in a descending column or down in an ascending column, or perhaps switching from one to another when necessary.

It doesn't seem hard to believe, really, and it would even be life-as-we-know-it; though, of course, we couldn't really be certain until we could figure out some way of actually exploring the Jupiter-ocean.

Although we have not yet explored any of the other outer giants as we have Jupiter (though several probes are *en route* to Saturn after having passed Jupiter), there seems no reason to doubt that what might be true for Jupiter might also be true for the others.

There might be four worlds, then, in the outer Solar System, that could be far richer in life than Earth.

Yet life on these outer planets would be ocean life, for planets that are largely made up of volatiles with a preponderance of hydrogen must be purely liquid. There is no way in which we can expect continents or even islands.

The life forms on the outer planets would, therefore, be very likely to be streamlined for getting rapidly through a medium more viscous than Earthly air and would, in consequence, be very apt to lack manipulative organs.

And even if they could manipulate the environment, could they develop the use of a convenient form of inanimate energy equivalent to our fire? (To be sure, there is no free oxygen on a planet like Jupiter, but there is free hydrogen, and oxygen-rich compounds might burn in a hydrogen atmosphere.)

Somehow, it seems rather likely that if life developed on the giant planets and evolved to the point of intelligence, it would be the intelligence of the dolphin rather than that of the human being. It would be an intelligence that might lead to a better way of life, but it would not involve the building of a technology based on ever more elaborate and sophisticated tools, with which the intelligent creature might directly manipulate the environment more and more subtly.

This would also be true of life developing, against the odds, in a possible water layer beneath the surface crust of Ganymede or Callisto.

In other words, there might be life on Jupiter and the other giant plants, even intelligent life – but it doesn't seem likely that there would be technological civilizations in our sense.

5 The stars

Substars

Having gone rather exhaustively through the Solar System, it would appear that although there may be life on several worlds other than Earth, even conceivably intelligent life, the chances are not high. Furthermore, the chances would seem to be virtually zero that a technological civilization exists, or could exist, anywhere in the Solar System but on Earth.

Nevertheless, the Solar System is by no means the entire Universe. Let us look elsewhere.

We might imagine life in open space in the form of concentrations of energy fields, or as animated clouds of dust and gas, but there is no hint of evidence that such a thing is possible. Until such evidence is forthcoming (and naturally the scientific mind is not closed to the possibility), we must assume that life is to be found only in association with solid worlds at temperatures less than those of the stars.

The only cool, solid worlds we know are the planetary and subplanetary bodies that circle our Sun, but we cannot assume from this that all such bodies in the Universe must be associated with stars.*

There may be clouds of dust and gas of considerably smaller mass than that from which our Solar System originated, and these may have ended by condensing into bodies much smaller than the Sun. If the bodies are sufficiently smaller than the Sun, say with only $\frac{1}{50}$ the mass or less, they would end by being insufficiently massive to ignite into nuclear fire. The surfaces of such bodies would remain cool and they would

---

* Our Sun, it is perhaps needless to say, is a star, and seems so different from all the rest only because it is so much closer to us.

resemble planets in their properties, except that they would follow independent motions through space and would not be circling a star.

All our experience teaches us that of any given type of astronomical body, the number increases as the size decreases. There are a greater number of small stars than of large ones, a greater number of small planets than large ones, a greater number of small satellites than large ones, and so on. Might we argue from that, that these substars, too small to ignite, are far greater in number than those similar bodies that are massive enough to ignite? At least one important astronomer, the American Harlow Shapley (1885–1972), has very strongly advanced the likelihood of the existence of such bodies.

Naturally, since they do not shine, they remain undetected and we are unaware of them. But if they exist, we might reason that there exist substars in space through an entire range of sizes from super-Jupiters to small asteroids. We might even suppose that the larger ones could have bodies considerably smaller than themselves circling them, much as there are bodies circling Jupiter and the other giant planets within our own Solar System.

The question is, though: Would life form on such substars?

So far I have suggested that the irreducible requirements for life (as we know it) are, first, a free liquid, preferably water, and, second, organic compounds. A third requirement, which ordinarily we take for granted, must be added, and that is energy. The energy is needed to build the organic compounds out of the small molecules present at the start, small molecules such as water, ammonia, and methane.

Where would the energy come from in those substars?

In the condensation of a cloud of dust and gas into a body of any size, the inward motion of the components of the cloud represents kinetic energy obtained from the gravitational field. When the motion stops, with collision and coalescences, the kinetic energy is turned into heat. The centre of every sizeable body is therefore hot. The temperature at the centre of the Earth, for instance, is estimated to be 5,000°C (9,000°F).

The larger the body and the more intense the gravitational field that formed it, the greater the kinetic energy, the greater the heat, and the higher the internal temperature. The temperature at the centre of Jupiter, for instance, is estimated to be 54,000°C (100,000°F).

It might be expected that this internal heat is a temporary phenomenon and that a planet would slowly but surely cool down. So it would, if there were no internal supply of energy to replace the heat as it leaked away into space.

In the case of Earth, for instance, the internal heat leaks away very slowly indeed, thanks to the excellent insulating effect of the outer layers of rock. At the same time, those outer layers contain small quantities of radioactive elements such as uranium and thorium, which, in their radioactive breakdown, liberate heat in large enough quantities to replace that which is lost. As a result, the Earth is not cooling off perceptibly, and though it has existed as a solid body for 4,600,000,000 years, its internal heat is still there.

In the case of Jupiter, there seem to be some nuclear reactions going on in the centre, some faint sparks of starlike behaviour, so that Jupiter actually radiates into space three times as much heat as it receives from the Sun.

This long-lasting internal heat would be more than ample to support life, if living things could tap it.

We could fantasize life as existing within the body of a planet where nearby pockets of heat might have served as the energy source to form and maintain it. There is, however, no evidence that life can exist anywhere but at or near the surface of a world, and until evidence to the contrary is obtained, we should consider surfaces only.

Suppose, then, we consider a substar no more massive than the Earth; or a body that massive that is circling a substar somewhat more massive than Jupiter but yielding no visible light.

Such an Earthlike body, whether free in space or circling a substar, would tend to be a world like Ganymede or Callisto. There would be internal heat, but, thanks to the insulating effect of the outer layers, very little would leak outwards to the surface; any more than Earth's internal heat leaks outwards to melt the snow of the polar regions and mitigate the frigidity of Earth's temperatures.

To be sure, on Earth there are local leaks of considerable magnitude, producing hot springs, geysers, and even volcanoes. We might imagine such things on Earth-sized substars as well. In addition, there could be energy derived from the lightning of thunderstorms. Still, whether such sporadic energy sources would meet the requirements for forming and maintaining life is questionable. There is also the point that a world without a major source of light from a nearby star may be unfit for the development of intelligence – a subject I will take up later in the book.

The Earth-sized substar would be composed of a much larger percentage of volatiles than Earth itself, since there would have been no nearby hot star to raise the temperature in surrounding space and make the collection of volatiles impossible. Therefore, again as on Ganymede and Callisto, we might imagine a world-girdling ocean, probably of water, kept liquid by internal heat, but covered by a thick crust of ice.

Substars still smaller than the Earth would have less internal heat and would be even more likely to be icy, have less in the way of sporadic sources of appreciable energy, have smaller oceans or none at all.

If a body were small enough to attract little or no volatile matter even at the low temperatures that would exist in the absence of a nearby star, it would be an asteroidal body of rock or metal or both.

What about substars that are larger than Earth and therefore possess greater and more intense reservoirs of internal heat? Such a larger body is bound to be Jupiterlike. A large substar is certain to be made up largely of volatile matter, particularly hydrogen and helium; and high internal heat will make the planet entirely liquid.

Heat can circulate much more freely through liquid by convection than through solids by slow conduction. We can expect ample heat at or near the surface in such large substars and the heat may remain ample for billions of years. However, again the most we can expect on a large substar is intelligent life of the dolphin variety – and no technological civilization.

In short, the formation of substars would rather resemble the formation of bodies in the outer Solar System, and we may expect no more of the former than of the latter.

For a technological civilization, we need a solid planet with both oceans and dry land, so that life as we know it can develop in the former and emerge on the latter. To form such a world there must be a nearby star to supply the heat that would drive away most of the volatile matter, but not all. The nearby star would also supply the necessary energy for the formation and maintenance of life in a copious and steady manner.

In that case, we must concentrate our attention on the stars. These, at least, we can see. We know they exist and need not simply assume the probability of their existence as in the case of the substars.

The Milky Way

If we turn to the stars and consider them as energy sources in the neighbourhood of which we may find life, possibly intelligence, and possibly even technological civilizations, our first impression may be heartening, for there seem to be a great many of them. Therefore, if we fail to find life in connection with one, we may do so in connection with another.

In fact, the stars may well have impressed the early, less sophisticated watchers of the sky as innumerable. Thus, according to the Biblical story, when the Lord wished to assure the patriarch Abraham that, despite his childlessness, he would be the ancestor of many people, this is how it is described:

'And he [God] brought him [Abraham] forth abroad, and said, "Look now towards heaven, and tell the stars, if thou be able to number them"; and he [God] said unto him [Abraham], "So shall thy seed be." '

Yet if God were promising Abraham that he would ultimately have as many descendants as there were stars in the
sky that he could see, God was not promising as much as
might be assumed.

The stars have been counted by later generations of astronomers who were less impressed with their innumerability. It
turns out the number of stars that can be seen with the unaided eye (assuming excellent vision) is, in total, about 6,000.

At any one time, of course, half the stars are below the horizon, and others, while present above the horizon, are so near
it as to be blotted out through light absorption by an unusually
great thickness of even clear air. It follows that on a cloudless,
moonless night, far from all man-made illumination, even a
person with excellent eyes cannot see more than about 2,500
stars at one time.

In the days when philosophers assumed all worlds were inhabited and when general statements to that effect were made,
it is not clear whether any particular philosopher truly understood the nature of stars.

Perhaps the first clear statement of the modern view was
that of Nicholas of Cusa (1401–1464), a cardinal of the Church,
who had particularly striking ideas for his time. He thought
that space was infinite and that there was no centre to the
Universe. He thought all things moved, including the Earth.
He also thought the stars were distant Suns, that they were
attended by planets as the Sun was, and that those planets
were inhabited.

Interesting, but we of the contemporary world are less
sanguine concerning habitability, and cannot accept in care-
free fashion the notion of life everywhere. We know there are
dead worlds, and we know that there are others, which while
possibly not dead, are not likely to bear more than simple
bacterialike forms of life. Why may there not be stars around
which only dead worlds orbit? Or around which no worlds
circle at all?

If it should turn out that habitability is associated with only
a small percentage of the stars (as life seems to be associated

with only a small percentage of the worlds of the Solar System), then it becomes important to determine whether there are stars other than those we happen to be able to see and if so, how many. After all, the greater the number of stars, the greater the chance of numerous life forms existing in space even if the chances for any one star are very low.

The natural assumption, of course, is that only those stars exist that can be seen. To be sure, some stars are so dim that excellent eyes can just barely make them out. Might it not seem natural to suppose that there are some that are fainter still and cannot be made out by even the best eyes?

Apparently, this seemed to occur to very few. Perhaps there was the unspoken feeling that God wouldn't create something too dim to be seen, since what purpose could such an object serve? To suppose that everything in the sky was there only because it affected human beings (the basis of astrological beliefs) seemed to argue against invisible bodies.

The English mathematician Thomas Digges (1543–1595) did espouse views like those of Nicholas of Cusa and in 1575 maintained not only infinite space, but an infinite number of stars spread evenly throughout it. Italian philosopher Giordano Bruno (1548–1600) also argued the same views, and did so in so undiplomatic and contentious a manner that he was finally burned at the stake in Rome for his heresies.

The argument over the matter ended in 1609, however, thanks to Galileo and his telescope. When Galileo turned his telescope on the sky, he immediately discovered that he saw more stars with his instrument than without it. Wherever he looked, he saw stars that could not be seen otherwise.

Without a telescope one saw six stars in the tiny little star group called the Pleiades. There were legends of a seventh that had dimmed and grown invisible. Galileo not only saw this seventh star easily once he clapped his telescope to his eyes, he saw thirty more stars in addition.

Even more important was what happened when he looked through his telescope at the Milky Way.

The Milky Way is a faint, lunimous fog that seems to form

a belt around the sky. In some ancient myths, it was pictured as a bridge connecting Heaven and Earth. To the Greeks it was sometimes seen as a spray of milk from the divine breast of the goddess Hera. A more materialistic way of looking at the Milky Way, prior to the invention of the telescope, was to suppose it was a belt of unformed star matter.

System of Stars Shaped Like a Coin

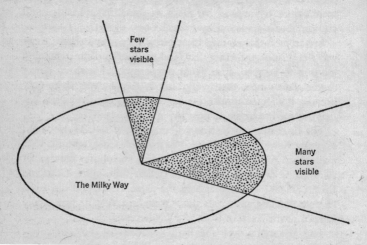

Few stars visible

Many stars visible

The Milky Way

When Galileo looked at the Milky Way, however, he saw it was made up of myriads of very faint stars. For the first time, a true notion of how numerous the stars actually were broke in on the consciousness of human beings. If God had granted Abraham telescopic vision, the assurance of innumerable descendants would have been formidable indeed.

The Milky Way, by its very existence, ran counter to Digges' view of an infinite number of stars spread evenly through infinite space. If that were so, then the telescope should reveal roughly equal numbers of stars in whatever direction it was pointed. As it was, it was clear that the stars did not stretch out equally in all directions, but that they made up a conglomerate with a definite shape to it.

The first to maintain this was the British scientist Thomas Wright (1711–1786). In 1750, he suggested that the system of stars might be shaped rather like a coin, with the Solar System near its centre. If we looked out towards the flat edges on either side, we saw relatively few stars before reaching the edge, beyond which there was none. If, on the other hand, we looked out along the long axis of the coin in any direction, the edge was so distant that the very numerous, very distant stars melted into dim milkiness.

The Milky Way, therefore, was the result of the vision following the long axis of the stellar system. In all other directions, the edge of the stellar system was comparatively nearby.

The whole stellar system can be called the Milky Way, but one usually goes back to the Greek phrase for it, which is *galaxias kylos* (*milky circle*). We call the stellar system the Galaxy.

The Galaxy

The shape of the Galaxy could be determined more accurately if one could count the number of stars visible in different parts of the sky, and then work out the shape that would yield those numbers. In 1784, William Herschel undertook the task.

To count all the stars all over the sky was, of course, an impractical undertaking, but Herschel realized it would be quite proper to be satisfied with sampling the sky. He chose 683 regions, well scattered over the sky, and counted the stars visible in his telescope in each one. He found that the number of stars per unit area of sky rose steadily as one approached the Milky Way, was maximal in the plane of the Milky Way, and minimal in the direction at right angles to that plane.

From the number of stars he could see in the various directions, Herschel even felt justified in making a rough estimate of the total number of stars in the Galaxy. He decided that it

contained 300 million stars, or 50,000 times as many as could
be seen with the unaided eye. What's more, he decided that the
Galaxy was five times as long in its long diameter as in its
short.

He suggested that the long diameter of the Galaxy was 800
times the distance between the Sun and the bright star Sirius.
At the time, the distance was not known, but we now know it
to be 8.63 light-years, where a light-year is the distance light
will travel in one year.*

Herschel's estimate, therefore, was that the Galaxy was
shaped like a grindstone, and was about 7,000 light-years
across its long diameter and 1,300 light-years across its short
diameter. Since the Milky Way seemed more or less equally
bright in all directions, the Sun was taken to be at or near the
centre of the Galaxy.

More than a century later, the task was undertaken again by
the Dutch astronomer Jacobus Cornelius Kapteyn (1851–1922).
He had the technique of photography at his disposal, which
made things a bit easier for him. He, too, ended with the decision that the Galaxy was grindstone-shaped with the Sun
near its centre. His estimate of the size of the Galaxy was
greater than Herschel's, however.

In 1906, he estimated the long diameter of the Galaxy to be
23,000 light-years and the short diameter to be 6,000 light-
years. By 1920, he had further raised the dimensions to 55,000
and 11,000 respectively. The final set of dimensions involved a
Galaxy with a volume 520 times that of Herschel's.

Even as Kapteyn was completing this survey of the Galaxy,
a totally new outlook had entered astronomical thinking.

It came to be recognized that the Milky Way was full of
clouds of dust and gas (like the one that had served as the

---

* Since light travels at the rate of 299,792 kilometres (186,282
miles) per second, a light-year is 9,460,000,000,000 kilometres
(5,878,500,000,000 miles) long. The distance to Sirius is therefore
82 trillion kilometres (50 trillion miles). It is simpler to use
light-years.

origin of our Solar System and, perhaps, of others) and that those clouds blocked vision. Thanks to those clouds, we could only see our own neighbourhood of the Galaxy and in that neighbourhood we were at the centre. Beyond the clouds, though, there might well be vast regions of stars we could not see.

Indeed, as new methods for estimating the distance of far-off star clusters were developed, it turned out that the Sun was not in or near the centre of the Galaxy at all, but was far off in the outskirts. The first to demonstrate this was Harlow Shapley, who in 1918 presented evidence leading to the belief that the centre of the Galaxy was a long distance away in the direction of the constellation Sagittarius, where, as it happens, the Milky Way is particularly thick and luminous. The actual centre was, however, hidden by dust clouds, as were the regions on the other side of the centre.

Through the 1920s, Shapley's suggestion was investigated and confirmed, and by 1930 the dimensions of the Galaxy were finally worked out, thanks to the labours of the Swiss-American astronomer Robert Julius Trumpler (1886–1956).

The Galaxy is more nearly lens shaped than grindstone shaped. That is, it is thickest at the centre and grows thinner towards its edges. It is 100,000 light-years across and the Sun is about 27,000 light-years from the centre, or roughly halfway from the centre towards one edge.

The thickness of the Galaxy is about 16,000 light-years at the centre and about 3,000 light-years at the position of the Sun. The Sun is located about halfway between the upper and lower edge of the Galaxy, which is why the Milky Way seems to cut the sky into two equal halves.

The Galaxy, as it is now known to be, is four times the volume of Kapteyn's largest estimate.

In a way, the Galaxy resembles an enormous Solar System. In the centre, playing the part of the Sun, is a spherical 'Galactic nucleus' with a diameter of 16,000 light-years. This makes up only a small portion of the total volume of the Galaxy, but it contains most of the stars. Around it are large numbers of stars that follow orbits about the Galactic nucleus as planets do around the Sun.

The Dutch astronomer Jan Henrick Oort (born 1900) was able to show in 1925 that the Sun was moving in a fairly circular orbit about the Galactic nucleus at a speed of about 250 kilometres (155 miles) per second. This speed is about 8.4 times the speed of the Earth moving around the Sun. The Sun and the whole Solar System revolve about the Galactic nucleus once every 200,000,000 years, so that in the course of its lifetime, so far, the Sun has completed perhaps twenty-five circuits about the Galactic nucleus.

From the speed of the Sun's progress about the Galactic nucleus, it is possible to calculate the gravitational attraction exerted upon it. From that and from the distance of the Sun from the Galactic centre, it is possible to calculate the mass of the Galactic nucleus and, roughly, of the entire Galaxy.

The mass of the Galaxy is certainly over 100 billion times that of our Sun, and some estimates place it as high as 200 billion times that of our Sun.

We might, quite arbitrarily, just in order to have a number to deal with, strike a point between the extremes and say (always subject to modification as better and more precise evidence is obtained) that the mass of the Galaxy is 160,000,000,000 times the mass of the Sun.

The mass of the Galaxy is distributed among three classes of objects. These are (1) stars, (2) nonluminous planetary bodies, and (3) clouds of dust and gas.

Although the nonluminous planetary bodies may conceivably be much more numerous than stars, each is so tiny compared to the stars that the total planetary mass must be small in comparison. Again, while the clouds of dust and gas take up enormous volumes, they are so rarefied that the total cloud mass must be small by comparison.

We can be sure that nearly all the mass of the Galaxy is in the form of stars. Although our own Solar System, for instance, contains but one Sun and innumerable planets, satellites, asteroids, comets, meteoroids, and dust particles circling it, that one Sun contains about 99.86 per cent of all the mass of the Solar System.

The stars of the Galaxy may not make up so overwhelming a percentage of the total mass as that, but it is fairly safe to suppose that they may make up 94 per cent of the mass of the Galaxy. In that case, the mass of the stars in the Galaxy is equal to 150,000,000,000 times the mass of the Sun.

Can that mass of stars be turned into the *number* of stars?

That depends on how representative the mass of the Sun is with respect to the mass of stars generally.

The Sun is a huge object compared to the Earth, or even compared to Jupiter. Its diameter is 1,392,000 kilometres (868,000 miles) or 110 times the diameter of the Earth. Its mass is two million trillion trillion kilograms, or 324,000 times the mass of the Earth. Nevertheless, it is not remarkable as stars go.

There are stars that are as much as 70 times as massive as the Sun and that shine a billion times as brightly. There are other stars that are only $\frac{1}{20}$ the mass of the Sun (and are therefore only 50 times the mass of Jupiter) and that flicker with a light only one-billionth that of the Sun.

Roughly speaking, one must conclude that the Sun is an average star, about equally distant from the extremes of giant size and brilliance on one end of the scale and pygmy size and dimness on the other end of the scale.

If the stars were equally distributed all along the mass scale and if the Sun were really average, then we would assume that there were 150 billion stars in the Galaxy.

As it happens, however, the smaller stars are more numerous than the larger ones, so that it is fair to estimate that the average star is about half the size of the Sun in mass. (There are small stars in which matter is very compressed and which are very dense, but their mass is not unusually high and they do not affect the average.)

If, then, the total mass of the stars in the Galaxy is 150 billion times the mass of the Sun, and the average star is 0.5 times the mass of the Sun, then it follows that there are some

300 billion stars in the Galaxy. This means that for each visible star in the sky, each one a member of the Galaxy, there are 50 million other stars in the Galaxy that we cannot see with our unaided eyes.

The other galaxies

Have we now come to an end? Are 300 billion stars all there are in the Universe? To put it another way, is the Galaxy all there is?

Suppose we consider two patches of luminosity in the sky that look like isolated regions of the Milky Way, and that are so far south in the sky as to be invisible to viewers in the North Temperate Zone. They were first described in 1521 by the chronicler accompanying Magellan's voyage of circumnavigation of the globe – so they are called the Large Magellanic Cloud and the Small Magellanic Cloud.

They were not studied in detail until John Herschel observed them from the astronomic observatory at the Cape of Good Hope in 1834 (the expedition that fuelled the Moon Hoax). Like the Milky Way, the Magellanic Clouds turned out to be assemblages of vast numbers of very dim stars because of their distance.

In the first decade of the twentieth century, the American astronomer Henrietta Swan Leavitt (1868–1921) studied certain variable stars in the Magellanic Clouds. By 1912, the use of these variable stars (called Cepheid variables because the first to be discovered was in the constellation Cepheus) made it possible to measure vast distances that could not be estimated in other ways.

The Large Magellanic Cloud turned out to be 170,000 light-years away and the Small Magellanic Cloud 200,000 light-years away. Both are well outside the Galaxy. Each is a galaxy in its own right.

They are not large, however. The Large Magellanic Cloud may include perhaps ten billion stars and the Small Magellanic Cloud only about two billion. Our Galaxy (which we may refer to as the Milky Way Galaxy if we wish to distinguish it from others) is twenty-five times as large as both Magellanic Clouds put together. We might consider the Magellanic Clouds as satellite galaxies of the Milky Way Galaxy.

Is this all, then?

A certain suspicion arose concerning a faint, fuzzy patch of cloudy matter in the constellation Andromeda; a patch of dim light called the Andromeda Nebula. Even the best telescopes could not make it separate into a conglomeration of dim stars. A natural conclusion was, therefore, that it was a glowing cloud of dust and gas.

Such glowing clouds were indeed known, but they did not glow of themselves. They glowed because there were stars within them. No visible stars could be seen within the Andromeda Nebula. The light from other luminous clouds when analysed, however, turned out to be completely different from starlight; whereas the light of the Andromeda Nebula was exactly like starlight.

Another alternative, then, was that the Andromeda Nebula was a conglomeration of stars, but one that was even more distant than the Magellanic Clouds, so that the individual stars could not be made out.

When Thomas Wright had first suggested in 1750 that the visible stars were collected into a flat disc, he theorized that there might be other such flat discs of stars at great distances from our own. This idea was taken up by the German philosopher Immanuel Kant (1724–1804) in 1755. Kant spoke of 'island universes'.

The notion did not catch on. Indeed, when Laplace developed his notion that the Solar System had formed out of a whirling cloud of dust and gas, he cited the Andromeda Nebula as an example of a cloud slowly whirling and contracting to form a sun and its attendant planets. That was the reason the theory was called the nebular hypothesis.

By the time the twentieth century opened, however, the old notion of Wright and Kant was gathering strength. Occasionally, stars did appear in the Andromeda Nebula, stars that were clearly 'novas'; that is, stars that suddenly brightened several magnitudes and then dimmed again. It was as though there were stars in the Andromeda Nebula that were ordinarily too dim to see under any circumstances because of their great distances, but that, upon briefly brightening with explosive violence, became just bright enough to make out.

There are such novas, now and then, among the stars of our own Galaxy, and by comparing their apparent brightness with the brightness of the very dim novas in the Andromeda Nebula, the distance of the Andromeda could be roughly worked out.

By 1917, the argument was settled. A new telescope with a 100-inch mirror had been installed on Mt Wilson, just north-east of Pasadena, California. It was the largest and best tele-scope that existed up to that time. The American astronomer Edwin Powell Hubble (1889–1953), using that telescope, was finally able to resolve the outskirts of the Andromeda Nebula into masses of very faint stars.

It was the 'Andromeda Galaxy' from that point on.

By the best modern methods of distance determination, it would appear that the Andromeda Galaxy is 2,200,000 light-years distant, eleven times as far away as the Magellanic Clouds. No wonder it was difficult to make out the individual stars.

The Andromeda Galaxy is no dwarf, however. It is perhaps twice as large as the Milky Way Galaxy and may contain up to 600 billion stars.

The Milky Way Galaxy, the Andromeda Galaxy, and the two Magellanic Clouds are bound together gravitationally. They form a 'galactic cluster' called the Local Group and are not the only members, either. There are some twenty mem-bers altogether. There is one, Maffei I, which is about 3,200,000 light-years away, and it is just about as large as the Milky Way. The remainder are all small galaxies, a couple with less than a million stars apiece.

There may be as many as 1.5 trillion stars in the Local Group altogether, but that isn't all there are either.

Beyond the Local Group, there are other galaxies, some single, some in small groups, some in gigantic clusters of thousands. Up to a billion galaxies can be detected by modern telescopes, stretching out to distances of a billion light-years.

Even that is not all there is. There is reason to think that, given good enough instruments, we could make observations as far as twelve billion light-years away before reaching an absolute limit beyond which observation is impossible. It may be that there are 100 billion galaxies, therefore, in the observable universe.

Just as the Sun is a star of intermediate size, the Milky Way Galaxy is one of intermediate size. There are galaxies with masses 100 times larger than that of the Milky Way Galaxy, and tiny galaxies with only a hundred-thousandth the mass of the Milky Way Galaxy.

Again, the small objects of a particular class greatly outnumber the large objects, and we might estimate rather roughly that there are on the average ten billion stars to a galaxy, so that the average galaxy is of the size of the Large Magellanic Cloud.

That would mean that in the observable universe, there are as many as 1,000,000,000,000,000,000,000 (a billion trillion) stars.

This one consideration alone makes it almost certain extraterrestrial intelligence exists. After all, the existence of intelligence is not a zero-probability matter, since *we* exist. And if it is merely a near-zero probability, considering that near-zero probability for each of a billion trillion stars makes it almost certain that somewhere among them intelligence and even technological civilizations exist.

If, for instance, the probability were only one in a billion that near a given star there existed a technological civilization, that would mean that in the Universe as a whole, a trillion different such civilizations would exist.

Let us move on, though, and see if there is any way we can put actual figures to the estimates; or, at least, the best figures we can.

In doing so, let us concentrate on our own Galaxy. If there are extraterrestrial civilizations in the Universe, those in our own Galaxy are clearly of greatest interest to us since they would be far closer to us than any others. And any figures we arrive at that are of interest in connection with our own Galaxy can always be easily converted into figures of significance for the others.

Begin with a figure that deals with our Galaxy and divide it by 30 and you will have the analogous figure for the average galaxy. Begin with a figure that deals with our Galaxy and multiply it by 3.3 billion and you have the analogous figure for the entire Universe.

We start then with a figure we have already mentioned:

1 The number of stars in our Galaxy = 300,000,000,000.

6 Planetary systems

Nebular hypothesis

The existence of the stars themselves, in no matter how huge a number, does not guarantee the existence of civilizations, or even of life, if *only* stars exist. The stars supply the necessary energy, but life must develop at a temperature compatible with the existence of the complex organic compounds that are the chemical basis of it.

This means that there must be a planet existing in the neighbourhood of the star. On that planet, warmed and, in general, energized by that star, life might conceivably exist.

We must therefore not consider stars, but planetary systems – of which our own Solar System is the only example that we know definitely and in detail.

Unfortunately, we cannot observe the neighbourhood of any star other than that of our own Sun with sufficient minuteness to be able to detect, directly, the presence of planets circling them.*

Does this defeat us at the start and make it impossible to come to any further conclusions as to the existence of extra-terrestrial intelligence?

Not necessarily. If we can determine how our own Solar System was formed, we might be able to draw conclusions as to the probability of the formation of other planetary systems.

For instance, the first theory of Solar System formation that many astronomers found attractive was Laplace's nebular hypothesis, which I mentioned earlier in the book. (Actually, something like it had been advanced by Kant in 1755, a half-century before Laplace.)

* There is tenuous and indirect evidence that they exist. This is something we will discuss later in the chapter.

If the Sun had formed out of the condensation of a spinning cloud of dust and gas (and we can see many such clouds in our Galaxy and in some other galaxies as well), it is reasonable to suppose that other stars formed in the same way.

Since our Sun, as it condensed, could be pictured as spinning faster and faster and losing rings of material from its equatorial region – one ring after another – thus forming the planets, other stars as they formed would do the same.

In that case, every star would have a planetary system.

We could not, however, come to that conclusion on the basis of the nebular hypothesis unless that theory of planetary formation could withstand close examination, and it didn't.

In 1857, Maxwell (who later worked out the kinetic theory of gases) was interested in reasoning out the constitution of Saturn's rings. He showed that if the rings were solid structures (as they seemed to be in the telescope) they would be broken up under the influence of Saturn's gravitational pull. It seemed, therefore, that they must consist of a large aggregate of relatively small particles, so thickly strewn as to seem solid when viewed from a great distance.

Maxwell's mathematical analysis turned out to be applicable to the ring of dust and gas supposedly shaken loose by the contracting nebula on its way to condensation into the Sun. It turned out that if Maxwell's mathematics was correct, it was difficult to see how such a ring would condense into a planet. It would at best form an asteroid belt.

An even more serious objection arose out of a consideration of angular momentum, which is the measure of the turning tendency of any isolated body or system of bodies.

Angular momentum depends on two things: the speed of each particle of matter as it rotates about an axis, or revolves about some distant body, or both; and the distance of each particle of matter from the centre of rotation. The total angular momentum of an isolated body can't vary in quantity, no matter what changes take place in the system. That is called the law of conservation of angular momentum. By this law, the velocity of spin must increase to make up for any decrease in distance, and vice versa.

A figure skater demonstrates the principle when she or he begins spinning with the arms outstretched, and then draws those arms in. At this condensation of the human body, so to speak, the rate of spin rapidly increases, and if the arms are then outstretched, it as rapidly slows down again.

When the rotating nebula gives off a ring of matter, this ring of matter cannot be more than a very small portion of the whole nebula. (This is obvious, since the ring condenses into a planet that is much smaller than the Sun.) Each bit of matter in the ring contains more angular momentum than a similar bit of matter from the main body of the nebula, because the ring comes off the equatorial belt where both the velocity of spin and the distance from the axis of rotation are highest. Nevertheless, the *total* angular momentum of the ring must be only a tiny fraction of the total angular momentum of all the rest of the vast nebula.

One would expect therefore that the Sun today, even after it has given off the matter required to form all the planets, would still retain much of the angular momentum of the original nebula. Its rate of spin should have accelerated so much as it shrank that it should today be rotating on its axis with violent speed.

Yet it doesn't. A point on the Sun's equator takes no less than twenty-six days to move once around the Sun's axis. Points north and south of the equator take even longer. This means that the Sun contains a surprisingly small amount of angular momentum.

The Sun, in fact, which contains 99.8 per cent of all the mass in the Solar System, possesses only 2 per cent of the angular momentum in the System. All the rest of the angular momentum is contained in the various small bodies that turn on their axes and swing around the Sun.

Fully 60 per cent of all the angular momentum in the Solar System is possessed by Jupiter and another 25 per cent by

Saturn. The two planets together, with only $\frac{1}{800}$ of the mass of the Sun, possess forty times as much angular momentum.

If all the spinning, revolving worlds of the Solar System were somehow to spiral into the Sun and add their angular momentum to the Sun's (as they would have to by the law of conservation of angular momentum), the Sun would spin on its axis in half a day.

There seemed no way in which so much angular momentum could be concentrated into the tiny rings peeling off the equatorial region of the spinning nebula and taken away from the nebula itself. Once this matter of angular momentum was clearly realized in the latter decades of the nineteenth century, the nebular hypothesis seemed to have received a death blow.

Stellar collisions

In the search for some explanation of the origin of the Solar System that would account for the peculiar distribution of angular momentum, astronomers veered away from evolutionary theories of planetary formation – that is, theories postulating slow but inexorable changes. They turned instead to catastrophic theories in which planets are formed by a sudden change that is not an inevitable part of the development, but an unexpected one.

In such theories, the original rotating nebula condenses smoothly to the Sun with no formation of planets. Rolling through space in solitary splendour, however, the Sun encounters a catastrophe that forms the planets *and* transfers angular momentum to them.

The first catastrophic theory was actually advanced in 1745, ten years before Kant had advanced the first version of the nebular hypothesis.* It was advanced by the French naturalist Georges Louis Leclerc de Buffon (1707–1788).

---

* Any naturalistic explanation of the formation of the Solar System could not greatly precede this. The strength of the belief in creationism (that is, the formation of the Universe in accordance with the description in Genesis i) was so strong up to that time

Buffon suggested that the planets, including Earth, had come into existence some 75,000 years before, as a result of a collision between the Sun and another large body, which he called a comet. (It was a time when the nature of comets was as yet unknown, but in which they were known to approach the Sun unusually closely.) Life, he thought, had then begun 35,000 years after Earth's formation. This conflicted with the general belief that God had created both the Earth and life less than 6,000 years earlier.

Buffon's notion, which lacked detail, receded into the background in view of the popularity of the nebular hypothesis. By 1880, however, when the nebular hypothesis was running into trouble over the matter of angular momentum, the catastrophe notion was revived.

The English astronomer Alexander William Bickerton (1842–1929) suggested that the Sun and another star passed close by each other. The gravitational influence of each body on the other pulled a stream of matter outwards. As the stars separated, the gravitational influence between them pulled that stream of matter sideways, imparting 'English' to it and giving it a great deal of angular momentum at the expense of the main portion of the bodies. From the streams of matter pulled out in the near-collision, the planets formed. Two solitary stars entered the state of near-collision; two stars with planetary systems emerged. It was a dramatic picture.

By 1880, a number of the galaxies had been made out in the telescopes of the time, and many of them had a glowing nucleus, together with spiral structures outside that nucleus. This was first noted in 1845 by the Irish astronomer William Parsons, Earl of Rosse (1800–1867).

At the time, it was not understood that these 'spiral nebulae' were vast and distant assemblages of stars and that our own Galaxy was one. They were thought to be small formations within our Galaxy, and Bickerton thought that they might

---

that to deviate from it would have put the deviator into serious jeopardy.

represent planetary systems in the process of formation, with the spiral arms representing the streams of matter pulled out of the central sun and given a strong curve that started them on their revolutions.

For the next fifty years, the catastrophic theory of planetary formation was popular with astronomers. The English astronomer James Hopwood Jeans (1877–1946) suggested that the stream of matter pulled out from the Sun was cigar shaped and that Jupiter and Saturn were formed from the fattest part of the stream, and that that was why they were so large. Jeans was a superb writer of popular science and his influence did more to impress the general public with this theory of the formation of the Solar System than did anything else.

Close analysis of the catastrophic theory, however, suggested difficulties. Could the streams of matter issuing from the Sun extend so far outwards as to give rise to the outer planets? Could the gravitational influence of the other star transfer enough angular momentum to the planets?

As a result, astronomer after astronomer attempted to modify the theory to make it more plausible. Some suggested an actual grazing collision rather than a mere pass-by. The American astronomer Henry Norris Russell (1877–1947) suggested that the Sun had been part of a two-star system, with the planets born of the other star so that they possessed its momentum.

Despite the difficulties, the catastrophic theories reigned supreme even into the 1930s, and this was a matter of crucial interest with respect to the thesis of extraterrestrial intelligence.

If the nebular hypothesis or any evolutionary theory of the Solar System were correct, then planets were formed as part of the normal development of a star and there were, essentially, as many planetary systems as there were stars. In that case, the chances of extraterrestrial intelligence might be very good.

The catastrophic theories, on the other hand, made planetary formation an accidental and not an inevitable thing. It depended on a sort of cosmic rape, on the fortuitous coming together of two stars.

As it happens, stars are so widely separated and move so slowly in comparison with the distance of separation that the chances of such a collision or near-collision are exceedingly small. During its entire lifetime, a star like the Sun has only one chance in five billion of closely approaching another star. In the entire lifetime of the Galaxy, there may have been only fifteen such close approaches outside the Galactic nucleus.

If any form of the catastrophic theory should correspond to reality, it would mean that there are very few planetary systems in the Galaxy, and the chance that any one of those few should harbour a civilization (excluding our own, of course) would be extraordinarily small.

Fortunately for the chances of extraterrestrial intelligence, however, the catastrophic theories proved less tenable with each decade.

Despite all the modifications introduced, there remained great difficulty in giving the planets sufficient angular momentum. Any mechanism that could be devised to provide it was all too apt to give them enough speed to cause them to escape from the Solar System altogether.

Then, in the 1920s, the English astronomer Arthur Stanley Eddington (1882–1944) worked out the internal temperature of the Sun (and of stars generally). The Sun's enormous gravitational field tends to compress its matter and pull it inwards, yet the Sun is gaseous throughout and has a density only about a quarter that of the Earth. Why does it not condense to much greater densities under the inexorable inward pull of gravity?

To Eddington, it seemed that the only thing that could counteract the inward pull of gravity would be the outward expansive force of internal heat. Eddington calculated the temperatures required to balance the gravitational in-pull and showed, quite convincingly, that the Sun's core had to be at temperatures of millions of degrees.

If then, as a result of a collision, or near collision, large amounts of matter were pulled out of the Sun, or of any star, that matter was going to be at much higher temperatures than had been thought. They would be so hot, the American astronomer Lyman Spitzer Jr (born 1914) pointed out in 1939, that there was no chance at all they would condense into planets. They would expand into thin gas and be gone.

Nebular hypothesis again

During the early 1940s, with the nebular hypothesis long dead and the catastrophic theory freshly killed, there was the uneasy feeling that *no* theories would explain the existence of the Solar System. It almost seemed that in sheer desperation one would have to believe that the Solar System was created by divine intervention after all, or that it didn't exist.

In 1944, however, the German astronomer Carl Friedrich von Weizsäcker (born 1912) returned to a form of the nebular hypothesis and introduced into it the kind of refinements that the developing state of knowledge had made possible since Laplace's day a century and a half before.*

According to the new version, the Sun did not contract and give off rings of gas in the process. Instead, the original nebula contracted, but left gas and dust behind as it did so. In this gas and dust, turbulences were set up – large whirlpools so to speak.

Where these whirlpools met, the particles in them collided and formed larger particles. At the very outskirts of the original nebula such particle formation may have resulted in a vast belt of small icy bodies, a few of which, now and then, alter their orbits under the influence of the gravitational attraction of nearby stars and enter the inner Solar System. There they make their appearance to us as comets.†

* A very similar theory was advanced simultaneously and independently by the Soviet astronomer Otto Yulyevich Schmidt (1891–1956), whose birthplace, as it happens, is only 130 kilometres (80 miles) from my own.
† Such a far-out belt of comets was first postulated by the American astronomer Fred Lawrence Whipple (born 1906) in 1963, long

Closer to the Sun, where the clouds of dust and gas are denser and more massive, larger bodies are formed – the planets.

The exact mechanism whereby the planets grew out of the turbulences wasn't easy to work out. Astronomers such as Kuiper and chemists such as the American Harold Clayton Urey (born 1893) improved on Weizsäcker's notions and suggested methods that apparently would allow the planets to grow satisfactorily.

There is still the matter of angular momentum, though. Why does the Sun turn so slowly that almost all the angular momentum is contained in the planets? What slowed the Sun?

Laplace understood the workings of gravitation, of course; no one better in his time, and few better since. In Laplace's time, however, there was no real understanding of the electromagnetic field that stars and planets also possess. Astronomers now know a great deal more about them, and these fields can be taken into account in any description of the origin of the Solar System.

The Swedish astronomer Hannes Olof Gösta Alfven (born 1908) worked out a detailed description of the manner in which the Sun gave off material in its early days (like the Solar wind of today, but stronger) and how this material, under the influence of the Sun's electromagnetic field, picked up angular momentum. It was the electromagnetic field that transferred angular momentum from the Sun to material outside the Sun and made it possible for the planets to be as far from the Sun as they are and to possess as much angular momentum as they do.

Now, a third of a century from the return of the nebular hypothesis, astronomers accept it with considerable confidence, along with its consequences.

In the new version of the nebular hypothesis, the outer

---

after Weizsäcker had first advanced his theory. Still later, Oort added detail and placed the belt very far from the Sun, a light-year or two away.

planets are not older than the inner planets; all the planets and the Sun itself are of the same age.

Furthermore, if the Sun and the planets formed out of the same whirlpools of dust and gas, all developing in the same process, then this is very likely the way in which any star like the Sun (and just possibly any star at all) develops. There should, in that case, be very many planetary systems in the Universe and just possibly as many planetary systems as there are stars.

The rotating stars

Is there any way we can check this suggestion of the universality of planetary systems? Theories are all very well, but if there is any physical evidence that can be gathered, however tenuous, so much the better.

Suppose we had evidence to show that planetary systems were few. We would have to suppose the Weizsäcker theory of star formation was wrong, or at least that it must be seriously modified. Perhaps the Sun formed in lonely splendour, and then passed through another cloud of dust and gas in space (there are plenty of such clouds) and collected some of it gravitationally. In that case, turbulences in the second cloud might finally form the planets, which would be younger than the Sun, perhaps a great deal younger.

This would be a return to a form of catastrophism, even though the passing of the Sun through a cloud of gas is not nearly so violent an event as the collision or near collision of two stars. It is still an accidental event and would necessarily result in relatively few planetary systems.

On the other hand, if it turned out that the evidence clearly indicated that a great many stars happened to have planets, then we could not possibly expect this to happen in any catastrophic way. Some version of the nebular hypothesis with the automatic or near-automatic formation of planets along with a star would have to be correct.

The trouble is, though, that we can't see whether any stars have planets in attendance. Even at the distance of the nearest star (Alpha Centauri, which is 4.3 light-years from us) there would be no way of actually seeing even a large planet the size of Jupiter or greater. Such a planet would be too small to see by the reflected light of its star. Even if a telescope were invented that could make out that dim flicker of reflected light, the nearness of the much greater light of its star would utterly drown it out.

We must give up hope of direct sighting then, at least for now, and resort to indirect means.

Consider our own Sun, which is a star that certainly has a planetary system. The remarkable thing about the Sun is that it rotates so slowly on its axis that ninety-eight per cent of the angular momentum of the system resides in the insignificant mass of its planets.

If angular momentum passed from the Sun to its planets when those planets were formed (by any mechanism), then it is reasonable to suppose that angular momentum might pass from any star to its planets. If, then, a star has a planetary system, we would expect it to spin on its axis relatively slowly; if it does not, we would expect it to spin relatively rapidly.

But how does one go about measuring the rate at which a star spins when even in our best telescopes it appears as only a point of light?

Actually, there is much that can be deduced from starlight even if the star itself is but a point of light. Starlight is a mixture of light of all wavelengths. The light can be spread out in order of wavelength from the short waves of violet light to the long waves of red light, and the result is a 'spectrum'. The instrument by which the spectrum is produced is the 'spectroscope'.

The spectrum was first demonstrated in the case of sunlight by Isaac Newton in 1665. In 1814, the German physicist Joseph von Fraunhofer (1787–1826) showed that the Solar spectrum was crossed by numerous dark lines, which, it was eventually realized, represented missing wavelengths. They were wavelengths of light that were absorbed by atoms in the Sun's atmosphere before they could reach the Earth.

In 1859, the German physicist Gustav Robert Kirchhoff (1824–1887) showed that the dark lines in the spectrum were 'fingerprints' of the various elements, since the atoms of each element emitted or absorbed particular wavelengths that the atoms of no other element emitted or absorbed. Not only could spectroscopy be used to analyse minerals on Earth, but it could be used to analyse the chemical makeup of the Sun.

Meanwhile, the art of spectroscopy had been refined to the point where the light of stars, though much dimmer than the light of the Sun, could also be spread out into spectra.

From the dark lines in the stellar spectra much could be worked out. If, for instance, the dark lines in the spectrum of a particular star were slightly displaced towards the red end, then the star would be receding from us at a speed that could be calculated from the extent of the displacement. If the dark lines were displaced towards the violet end of the spectrum, the star would be approaching us.

The significance of this 'red shift' or 'violet shift' was quite evident from work that had been done on sound waves in 1842 by the Austrian physicist Christian Johann Doppler (1803–1853) and then applied to light waves in 1848 by the French physicist Armand Hippolyte Louis Fizeau (1819–1896).

Suppose, now, that a star is rotating and that it is so situated in space that neither of its poles is facing us, but that each pole is located at or near the sides of the star as we view it. In that case, at one side of the star between the poles the surface is coming towards us, and on the opposite side it is receding from us. The light from one side causes the dark lines to shift slightly towards the violet, the light from the other causes them to shift slightly towards the red. The dark lines, shifting perforce in both directions, grow wider than normal. The more rapidly the star rotates, the wider the dark lines in the spectrum.

This was first suggested in 1877 by the English astronomer

William de Wiveleslie Abney (1843–1920); and the first actual discovery of broad lines produced by rotation came in 1909 through the work of the American astronomer Frank Schlesinger (1871–1943). It was only in the mid-1920s, however, that studies on the rotation of stars began to be common and the Russian-American astronomer Otto Struve (1897–1963) was particularly active here.

It was indeed found that some stars do rotate slowly. A spot on the Sun's equator travels only about two kilometres (1¼ miles) per second as the Sun makes its slow rotation on its axis, and many stars rotate with that equatorial speed or not very much more. On the other hand, some stars whirl so rapidly on their axis as to attain equatorial speeds of anywhere from 250 to 500 kilometres (165 to 330 miles) per second.

It is tempting to assume that the slow-rotators have planets and have lost angular momentum to them, while the fast-rotators do not have planets and have retained all, or almost all, their original angular momentum.

That is not all that can be learned in this way, however. When stellar spectra were first studied, it was clear that while some had spectra resembling that of the Sun, others did not. In fact, stellar spectra differed from each other widely and, as early as 1867, Secchi (the astronomer who had anticipated Schiaparelli's discovery of the Martian canals) suggested that the spectra be divided into classes.

This was done, and eventually the various attempts to label the classes ended in the spectra being listed as O, B, A, F, G, K, and M, with O, representing the most massive, the hottest, and the most luminous stars known; B was next, A next, and so on down to M, which included the least massive, the coolest, and the dimmest stars. Our Sun is of spectral class G and is thus intermediate in the list.

As stellar spectra were more and more closely studied, each spectral class could be divided into ten subclasses: B0, B1 ... B9; A0, A1 ... A9; and so on. Our Sun is of spectral class G2.

The American astronomer Christian Thomas Elvey (born 1899), working with Struve, found by 1931 that the more massive a star, the more liable it was to be a fast-rotator. The stars of spectral classes O, B, and A, together with the larger F-stars, from F0 to F2, were very likely to be fast-rotators.

The stars of spectral classes F2–F9, G, K, and M were virtually all slow-rotators.

Half the spectral classes, then, are fast-rotators and half are slow-rotators, but that doesn't translate into an equal division of stars. The smaller stars are more numerous than the larger ones, so that there are more stars, by far, that are spectral class G or smaller than are spectral class F or larger. In fact, only 7 per cent of all the stars are included in spectral classes 0 to F2.

In other words, there are not more than 7 per cent of the stars that are fast-rotators and fully 93 per cent of the stars that are slow-rotators. This would make it seem that at least 93 per cent of the stars have planetary systems.

In fact, we might not even be truly able to eliminate the 7 per cent of the fast-rotators. They happen to include the particularly massive stars, which are likely to have a much higher total angular momentum to begin with than smaller stars would have. They might have enough angular momentum left to spin rapidly even after they had lost some to their planets.

Or – the loss of angular momentum to the planets may take time and as we shall see, the really massive stars are all young stars. It may be that they haven't yet had time to transfer the angular momentum.

From the data on stellar rotation, then, it seems fair to conclude that at least 93 per cent – and possibly 100 per cent – of stars have planetary systems.

The wobbling stars

So far, so good, but we must admit that stars may be fast-rotators or slow-rotators for reasons that have nothing to do with planets. Some stars may simply form from clouds that have more angular momentum to begin with – or less.

Can we therefore look for other types of evidence?

We can, if we stop to consider that when two bodies attract each other gravitationally, the attraction is two-way. The Sun attracts Jupiter, but Jupiter also attracts the Sun.

If two bodies, attracting each other gravitationally, were exactly equal in mass, neither would rotate about the other, properly speaking. Contributing equally to the gravitational interaction, they would each circle around a point exactly mid-way between the two. This point around which they would circle is the 'centre of gravity'.

If the two bodies were unequal in mass, the more massive body would be less affected by the attraction and would move less. If the more massive body is twice the mass of the less massive, the centre of gravity would be twice as close to the centre of the more massive body as to the centre of the less massive body. Suppose we consider the Moon and the Earth. The Moon is usually considered as revolving about the Earth, but it doesn't revolve about the Earth's centre. Both it and the Earth revolve about a centre of gravity that always lies between Earth's centre and the Moon's centre.

As it happens, the Earth is 81 times as massive as the Moon, so the centre of gravity has to be 81 times as close to the centre of the Earth as to the centre of the Moon. The centre of gravity of the Earth-Moon system is 4,750 kilometres (2,950 miles) from the Earth's centre. It is 348,750 kilometres (239,000 miles), 81 times as far, from the Moon's centre.

The centre of gravity of the Earth-Moon system is so close to the Earth's centre that it is 1,600 kilometres (1,000 miles) under the Earth's surface. Under the circumstances, it is certainly reasonable to consider the Moon as revolving about the Earth; it is, after all, revolving about a point inside the Earth.

The centre of the Earth also moves in a small circle about that centre of gravity once every 27⅓ days. If the Moon weren't there, the Earth would move around the Sun in a smooth path. Because of the presence of the Moon, the Earth makes a small wave 27⅓ days long in its path about the Sun – twelve and a fraction of these waves through each complete turn. The wobble of the Earth could, in theory, be measured from out in space, and from it the presence of the Moon and perhaps its distance and size could be worked out even if, for some reason, it could not be directly seen.

This is true of Jupiter and the Sun, too. The Sun is 1,050 times as massive as Jupiter, so the centre of gravity of the Sun-Jupiter system should be 1,050 times as close to the Sun's centre as it is to Jupiter's centre. Knowing the distance between the two centres, it turns out that the centre of gravity is 740,000 kilometres (460,000 miles) from the centre of the Sun. This means that the centre of gravity is 45,000 kilometres (28,000 miles) *outside* the Sun's surface.

The centre of the Sun circles this centre of gravity every twelve years. The Sun, in its smooth progress about the centre of the Galaxy, wobbles slightly, moving first to one side of its path, then to the other.

If only the Sun and Jupiter existed, an observer from a post in space, from which it was too far to see Jupiter directly, might deduce the presence of Jupiter from the Sun's wobble.

Actually, the Sun also possesses three other large planets: Saturn, Uranus, and Neptune, each of which has a centre of gravity with the Sun, though not one that is ever as far from the Sun's centre as Jupiter's. This makes the Sun's wobble a rather complicated one and that much harder to interpret.

Then, too, if the observer were as far away as one of the nearest stars, the Sun's wobble would be too small to measure accurately or even, perhaps, to detect.

Would it be possible to turn the tables? Could we look at some other star and detect a wobble in *its* path and from that deduce that *it* had a planet or planets?

Undoubtedly in some cases, for it was done as long ago as 1844.

In that year, the German astronomer Friedrich Wilhelm Bessel (1784–1846) noted a wobble in the motion of the bright star Sirius. From that wobble, he deduced the presence of an unseen companion that had $\frac{2}{5}$ the mass of Sirius.

As it happens, we now know that Sirius is 2.5 times as massive as our Sun. The companion, therefore, has just about the mass of our Sun. So it is not a planet, actually, but a full-sized star that is dim and hard to see because it happens to be very compact.*

To find a companion star is easy by comparison to finding a companion planet, however. A planet is so small in mass compared with the star it circles that the centre of gravity between itself and the star is that much closer to the centre of the star. The star therefore makes a very tiny wobble indeed.

Can such a wobble ever be measured?

Possibly, if the conditions are right.

First, the star must be as close to us as possible, so that the wobble is as large in appearance as possible.

Second, the star must be a small one, certainly smaller than our Sun, so that its mass predominates as little as possible. The centre of gravity is then comparatively far from the star's centre and this star makes a comparatively large wobble.

Third, the star must have a large planet, at least as large as Jupiter, so that the planetary mass will be large enough to drag the centre of gravity far enough away from the small star it circles to force a comparatively large wobble on the star.

This triple requirement of a nearby small star with a large planet cuts down the possibilities enormously. If the chance of planetary formation is small, then it would be too much to ask of coincidence that a planetary system should just happen to exist around a small nearby star, and that the planetary system should just happen to include a planet at least as large as Jupiter.

On the other hand, if we search small nearby stars and *do* happen to find evidence of an accompanying planet around at least one of them, then, in order not to force ourselves to accept a highly unlikely coincidence, we must consider that planetary systems are very common, perhaps even universal.

* These massive, but small and very dense stars, and others even more massive, smaller, and denser, are of no matter to us in this book and they will never be more than alluded to. If you are curious about them, you will find a complete discussion in my book *The Collapsing Universe* (Walker, 1977).

Attempts to determine the presence or absence of such wobbles in the motions of stars were conducted at Swarthmore College under the guidance of the Dutch-American astronomer Peter Van de Kamp (born 1901).

The Danish-American astronomer Kaj Aage Gunnar Strand (born 1907), working under Van de Kamp, detected a tiny wobble in the motion of one of the stars of the 61 Cygni two-star system, and deduced the presence of a companion body circling it, one that was much too small in mass to be a star. It was massive enough to be a large planet, however, one that was eight times as massive as Jupiter. The discovery was announced in 1943.

Since then a similar wobble was discovered in connection with Barnard's star, a small star only six light-years away. In its case, the wobble may indicate the presence of two planets, one as massive as Jupiter, orbiting in 11.5 years, and one as massive as Saturn, orbiting in 20 or 25 years. Other nearby stars, such as Ross *614* and Lalande *21185* have also shown wobbles that seem to indicate the presence of large planets.

In short, we have discovered not one but half a dozen small, nearby stars that may have large planets. Under the circumstances (and it must be admitted that the observations are so close to the limit of what can be seen that not all astronomers are ready to accept the conclusion without cautious reservations) it would seem that we must conclude that planetary systems are very common and that all the slow-rotating stars, at least, have them.

Let us be conservative and confine the planetary systems only to the slow-rotating stars, which make up ninety-three per cent of the whole.

In that case we get our second figure:

2 The number of planetary systems in our Galaxy = 280,000,000,000.

7 Sunlike stars

Giant stars

The fact that, according to our conclusions in the previous chapter, there is an enormous number of planetary systems in our Galaxy does not, *in itself*, mean that life is rampant.

Different stars may not be equally suitable as incubators of life on their planets and the next step is, therefore, to consider this possibility and to determine (if we can) which stars are suitable, and how many such suitable stars there might be.

If it turns out that the requirements for a suitable star are exceedingly numerous and complex, it may be that virtually no stars are suitable, and all those planetary systems might as well not be there, at least as far as extraterrestrial intelligence is concerned.

Such extreme pessimism is, however, unnecessary, for we begin with two statements, one of which is absolutely certain.

The certain statement is that our Sun is adequate as an incubator of life, so it is therefore possible for a star to be suitable. The second statement, somewhat less than completely certain but so near to certainty that no astronomer doubts the fact, is that the Sun is not a particularly unusual star. If the Sun is suitable, many stars should be.

Let us begin by asking how stars might differ.

The most obvious point of difference, one that was recognized as soon as inquisitive eyes turned upwards towards the night sky, is that the stars differ in brightness.

This difference, of course, may be due entirely to differences in distance. If all stars were equally bright when viewed at a given distance (if all, in other words, were of equal 'luminosity'), then those that were nearer to us, in actual fact, would be brighter in appearance than those that were farther from us.

Once the distances of the stars were worked out (the first to accomplish the task, in 1838, was Bessel, who six years later discovered Sirius's companion star) it turned out that the apparent brightnesses were not entirely due to different distances. Some stars are intrinsically more luminous than others.

Some stars are more massive than other stars, too, but mass and luminosity go hand in hand. As Eddington showed in the 1920s, a more massive star *had* to be more luminous. A more massive star had a more intense gravitational field and, in order to keep it from collapsing, the temperature at its centre had to be higher. A higher central temperature produced a greater flood of energy pouring out of the star in all directions, and its surface was both hotter and more luminous.*

What is more, luminosity goes up more rapidly than mass. If Star A is two times as massive as Star B, then Star A has a greater tendency to collapse in on itself because its gravitational field is greater. To withstand the greater gravitational field of Star A, the centre of that star must be much hotter; sufficiently hotter to make Star A ten times as luminous as Star B.

The most massive stars known are some 70 times the mass of the Sun, but they are 6 million times as luminous. On the other hand, a star with only $\frac{1}{16}$ the mass of the Sun (65 times the mass of Jupiter) might be just massive enough to glow a dull red heat, and it would only be one-millionth as luminous as the Sun.

What would it be like for a planet circling a star at such extremes?

Suppose, for instance, Earth were circling a star seventy times as massive as the Sun.

Of course, if Earth were circling this giant star at the same distance at which it circles the Sun, the star would appear forty times as wide in the sky as the Sun does to us, and it

---

* A very massive star may radiate so much of its energy in the invisible ultraviolet region that it will seem less luminous (to the human eye) than one might expect it to be.

would deliver six million times as much light and heat. The Earth would be a ball of red-hot rock.

We can easily imagine, however, that every star has a shell around it at some distance, within which a planet could circle and be heated by the star it circles to Earthlike standards of comfort. For a large star this shell, or 'ecosphere',* would be farther away than for a small star. In the case of the seventy-times–Sun giant, the ecosphere would be at a distance of hundreds of billions of kilometres from the star.

Suppose, then, that the Earth circled the giant star at a distance of 366 billion kilometres (227 billion miles). This would be a distance 2,450 times the distance of the Earth from the Sun and sixty-two times as far as Pluto is from the Sun. At such a distance it would take 14,500 years for the Earth to revolve about the star.

From that magnificent distance, the giant star would seem very small, so small that it would show no visible disc, but would shine merely like a star, but not like the stars we see. It would be extraordinarily bright because its temperature would be so much higher than that of the Sun (50,000°C as compared to a mere 6,000°C) that even though the giant star was so distant and so small in appearance, it would deliver as much light and heat to the distant planet as the Sun does to Earth.

To be sure, the giant star's temperature alters the nature of its radiation. At the distance we have imagined for Earth, the star would deliver the same total amount of energy that the Sun delivers now, but a much larger fraction of the giant star's energy would be in the form of ultraviolet light and x-rays, and a much smaller fraction would be visible light.

Human eyes are adapted to respond to visible light so that the light of the giant star would seem dimmer than that of the Sun. On the other hand, the flood of ultraviolet and x-rays would be deadly to Earth life.

Yet perhaps this is not a fatal objection. The Earth's atmosphere protects us against the energetic radiation of our Sun and we can imagine Earth moved still farther from the giant

* *Eco* – is from the Greek for *home* or *habitat*.

star. The decline in total radiation and the amount stopped by a possibly thicker atmosphere might then be suitable for the development of life at the price of somewhat lower planetary temperatures than we are used to.

There is, however, a more vital objection to the giant star, one that can't be countered by adjusting the planetary place within the ecosphere or by fiddling with the planetary atmosphere.

A star is not an adequate incubator for life throughout its existence. It cannot supply the energy necessary for life, for instance, while it is condensing and forming out of the primal nebula. It must first condense to the point where the nuclear fires start at the centre and it begins to radiate light. Eventually, the condensation reaches a stable stage and the radiation, having reached some maximum figure, remains there.

The star is then said to have entered the 'main sequence'. (It is called the main sequence because about ninety-eight per cent of the stars we can see are in that state, forming a sequence from the most massive to the least massive.)

While on the main sequence, a star's radiation is steady and reliable and, like our Sun, it could conceivably serve as an incubator for life.

The star's radiation depends, however, on the energy that develops as the hydrogen at its core is converted through processes of nuclear fusion into helium. At some critical point, when a large part of the hydrogen has been used up, the process begins to falter. The helium, accumulating in the core, renders the core more and more massive. It shrinks and condenses, and its temperature goes up to the point where helium fuses to form still more complicated nuclei.

At this point, the star develops enough heat to cause itself to expand against the pull of its own gravity, whereas till then, while it was on the main sequence, the inward pull of gravity and the outward push of temperature had remained in balance.

As the star now expands it leaves the main sequence and becomes relatively enormous in extent. Because of the expansion, the surface of the star cools and becomes merely red hot, though the total radiation from its now-vast surface is much greater than it had been before. The star is a red giant.

Once a star leaves the main sequence, what follows is hectic. It remains a red giant for several hundred million years (only a short time on the astronomical scale), while what is left of the hydrogen is consumed and while the core grows hotter and hotter. Finally there is a collapse, when the energy developed by nuclear fusion at the centre fails as all possible nuclear fuels are used up and the star can no longer be kept distended against its own gravity.

If the star is massive enough, the collapse is preceded by a cataclysmic explosion – a supernova. The more massive the star, the more drastic the explosion. What is left of the star then shrinks into a relatively tiny and very dense ball.*

As far as life is concerned, though, the details of what happens after the star leaves the main sequence are irrelevant. As the star begins to expand towards the red giant stage, its total radiation increases dramatically. Any planet that till then had been in a position to receive radiation in quantities consistent with the formation and maintenance of life would now receive far too much. Any life present would be baked to death. (In extreme cases, the planet itself would melt and evaporate.)

We can state, therefore, that as a general, and possibly inviolable, rule, a star can serve as an incubator of life only while it is on the main sequence.

Fortunately, a star can remain on the main sequence for a long time. Our Sun, for instance, may remain on the main sequence for a total period equal to 12 or 13 billion years. Although it has been shining now, in much its present fashion, for some 5 billion years, its life as a main sequence star is not yet half over.†

---

* For details on all this, see my book, *The Collapsing Universe*.
† The Sun will gradually grow warmer as it ages and by its final billion years on the main sequence, life may not be possible on Earth. When the Sun expands to a red giant, it will engulf the orbits of Mercury and Venus, and though Earth will probably

A star that is more massive than the Sun and therefore must counter the in-pulling effect of a stronger gravitational field, must develop higher temperatures at the centre to counter gravitational contraction and, to do that, must fuse hydrogen at a greater rate. To be sure, a star more massive that the Sun possesses more hydrogen to begin with, but the increase in the rate of fusion is greater than the increase in the hydrogen supply.

The more massive the star, then, the more rapidly it consumes its admittedly greater hydrogen supply, and the shorter its stay on the main sequence.

A monster star that is seventy times as massive as the Sun must consume its hydrogen at so fearsome a rate to remain expanded under the pull of its monster gravity that its life on the main sequence may be only 500,000 years or less. Indeed, that is why we observe no stars with really large masses. Even if gigantic stars formed, the temperatures they would develop would blow them up virtually at once.

Of course, even 500,000 years is a long time as far as human experience is concerned. Human written history has, at best, existed for only one-hundredth that period.

Intelligent life, however, did not come upon the Earth at its very beginning, but only as the result of a long course of evolution. If our Sun had only shone as it does now for 500,000 years after the formation of the Earth, and had then left the main sequence, it is highly doubtful if there would have been time for even the simplest protolife to form in Earth's oceans.

In fact, judging from the experience of Earth, it takes some five billion years of planetary existence for life to develop to the point of complexity where a civilization can be established.

We can't, of course, be sure how typical Earth's case is of the Universe as a whole. It may be that evolution has, for some trivial reason or other, been extraordinarily slow on Earth, and

remain outside the Sun's swollen sphere, it will at best be a red-hot ball of rock.

that on other planets much less time has been required for the evolution of intelligence. It may, on the other hand, be that evolution on Earth has, for some trivial reason or other, been extraordinarily rapid, and that on other planets much more time is required for the evolution of intelligence.

There is no way, at the moment, in which we can tell whether either alternative is true. We have no recourse but to adhere to 'the principle of mediocrity' and to assume that this one case we know of – that of Earth – is not atypical, but is about average in its nature.

We must, therefore, cling to a five-billion-year lifetime on the main sequence as an essential minimum for the development of civilization.

A star that is 1.4 times as massive as the Sun and is of spectral class F2 remains on the main sequence for five billion years, and we can therefore come to the conclusion that any star more massive than 1.4 times the mass of the Sun will not serve as an appropriate incubator for life. There may indeed be life on a planet circling such a too-massive star, but the chance that it will exist long enough to reach the appropriate pitch of complexity to produce an extraterrestrial civilization is small enough to ignore.

This means that the bright stars we see in the sky, which are (at least, most of them are) considerably more massive than the Sun, are unsuitable incubators. Sirius, for instance, will remain on the main sequence for 500 million years altogether, Rigel for only 400 million years. We can ignore such stars.

As it happens, however, it is precisely these massive short-lived stars that are fast-rotators and were therefore not included by me in the number of stars possessing a planetary system. Their exclusion from further consideration is thus doubly justified.

Midget stars

Let's try the other extreme, now, and consider a star with $\frac{1}{16}$ the mass of the Sun and one-millionth the luminosity. (Any object less massive than that would probably not be enough to ignite the nuclear fires to the centre and would not, therefore, be a true star.)

A midget with $\frac{1}{16}$ the mass of the Sun would be sixty-five times as massive as the planet Jupiter, but would surely be much more dense and might not be much larger than Jupiter in size. It might perhaps be 150,000 kilometres (93,000 miles) in diameter.

Next, suppose that Earth were 300,000 kilometres (186,000 miles) from the centre of such a star and therefore circling it at a height of 150,000 kilometres (93,000 miles) above its surface. Earth would circle that star every 1.1 hours.

Earth would receive as much total energy from that very nearby midget star as the Earth now does from the Sun. The fact that the midget star would be barely red hot would be made up for by the fact that from the distance of the planet its apparent size would be 3,000 times that of the Sun as we see it from Earth.

To be sure, the nature of the energy received from the midget star would be different from that of the Sun. The midget star would deliver virtually no ultraviolet radiation and, in fact, very little visible light. Most of its energy would be in the form of infrared light.

This would be very inconvenient from our standpoint. To our own eyes, everything would seem very dim and unpleasantly deep red in colour. We could imagine, however, that life on such a planet would have developed a sense of sight that would be sensitive to red and infrared, and perhaps see sections of it in different colours. To such life, the light might well appear white and sufficiently bright.

Red and infrared are less intensively energetic than the remainder of the visible light spectrum, and there would be many chemical reactions that yellow, green, or blue light could initiate that red and infrared could not. However, life is not based on photochemical reactions, except for photosynthesis and that is initiated by red light. No doubt we would not have to stretch matters intolerably to imagine life on such a world – so far.

Let us, however, take up a new issue:

The gravitational field of any object decreases in intensity with the square of the distance. If distance is doubled, the intensity falls to $\frac{1}{4}$ of what it was; if the distance is tripled, it falls to $\frac{1}{9}$ and so on.

This affects the manner in which the Moon and the Earth attract each other.

The average distance between the centre of the Moon and the centre of the Earth is 384,390 kilometres (238,860 miles). This varies somewhat as the Moon moves about its orbit, but that doesn't affect the line of argument.

Not all parts of the Earth are, however, at the same distance from the Moon. When the centre of the Earth is at its average distance from the centre of the Moon, the surface of the Earth that directly faces the Moon is 6,356 kilometres (3,950 miles) closer. The surface of the Earth that faces directly away from the Moon is 6,356 kilometres (3,950 miles) farther.

This means that while the surface of the Earth directly facing the Moon is at a distance of 378,034 kilometres (234,910 miles) from the Moon's centre, the surface of the Earth facing directly away from the Moon is at a distance of 390,746 kilometres (242,810 miles) from the Moon's centre.

If the distance of the Earth's near side from the Moon's centre is set at one, the distance of the Earth's far side is 1.0336. This difference, only 3.36 per cent of the total distance from the Moon, does not seem like much. However, the gravitational pull of the Moon falls off over that small distance by an amount equal to $\frac{1}{1.0336^2}$ and is only 0.936 at the far side compared with 1.000 at the near side.

The result of this difference in the Moon's pull at the near and far sides of the Earth is that the Earth is stretched in the direction of the Moon. The near surface is pulled towards the Moon more forcibly than the centre is, and the centre is pulled towards the Moon more forcibly than the far surface is. Both the near and far surface bulge, the former towards the Moon, the latter away from the Moon.

It is a matter of a small bulge only, half a metre or so. Still, as the Earth rotates, each part of its solid matter bulges up when it turns towards the side facing the Moon, reaching its greatest height when it passes under the Moon, then settling back. The solid matter bulges as it turns towards the side away from the Moon, reaching another peak when it is directly opposite the position of the Moon, then receding.

The water of the ocean bulges up also, to a greater extent than the solid land does. This means that as the Earth turns, the land surface passes through the higher bulge of water and, as it does so, the water creeps up the shore and then back down. It does so as it passes through both bulges of water, one on the side facing the Moon and one on the side away from it. This means the water rises and falls along the shore twice a day; or, we can say, there are two 'tides' a day.

Because this difference in gravitational pull causes the tides, it is referred to as a tidal effect.

Naturally, the Earth also exerts a tidal effect on the Moon. Since the Moon is smaller than the Earth, the Moon's diameter being 3,476 kilometres (2,160 miles) as compared with Earth's diameter of 12,713 kilometres (7,900 miles), the drop in gravitational pull across the Moon is smaller than the drop across the Earth.

The width of the Moon is only 0.90 per cent of the total distance between the Earth and the Moon, so that the gravitational pull on the far side is 98.2 per cent of the force on the near side. The tidal effect on the Moon would be, in this respect, only 0.29 times what it is for the Earth, *but* the Earth's gravitational field is eighty-one times that of the Moon, since the Earth is eighty-one times as massive as the Moon. If we multiply 0.29 by 81, we find that the tidal force of the Earth on the Moon is 23.5 times that of the Moon on the Earth.

Does this difference matter? Yes, it does.

As the Earth turns and bulges, the internal friction of the rock as it lifts up and settles down, and the friction of the water moving up the shore and back, consumes some of the energy of Earth's rotation and turns it into heat. As a result, tidal action is slowing the Earth's rotation. However, the Earth is so massive and the energy of its turning is so huge that the Earth's rotation is slowing very slowly indeed. The length of the day is increasing by one second every 100,000 years.*

This isn't much on the human time scale, but if the Earth has been in existence for five billion years and this rate of day lengthening has been constant throughout, the day has lengthened a total of 50,000 seconds or nearly fourteen hours. When the Earth was created, it may have been rotating on its axis in only ten hours – or less, if the tides were more important in early geologic times than they are now, as they well might have been.

What about Earth's tidal effect on the Moon?

The Moon has a smaller mass and therefore, very likely, a smaller rotational energy to begin with. Furthermore, the tidal effect on the Moon is 23.5 times that on the Earth. The stronger effect, working on the smaller mass, has a greater slowing effect. As a result, the Moon's rotational period has slowed until it is now equal to exactly one revolution about the Earth. Under those conditions, the same side of the Moon always faces the Earth, the tidal bulge is always in the same spot on its surface, so that different parts of its body no longer have to heave up and settle back as it turns. There is no further slowing (at least as far as Earth's tidal effect on the Moon is concerned) and the Moon's rotational period is now stable.

* The slowing of the rotation means a loss of angular momentum that by the law of conservation of angular momentum can't really be lost. What happens is that the Moon is slowly moving farther away from the Earth and so is the centre of gravity of the Earth–Moon system. What the Earth loses in the angular momentum of rotation, it gains in the angular momentum of a larger swing about a more distant centre of gravity.

As a result of tidal effect, small bodies would always be expected to turn only one face to the large bodies they circle. (This was first suggested by Kant in 1754.) Not only does the Moon turn only one face to the Earth, the two Martian satellites turn only one face to Mars, the five innermost satellites of Jupiter turn only one face to Jupiter, and so on.

In that case, though, why doesn't the Earth turn only one side towards the Sun?

Consider what would happen if the Moon receded from the Earth. As it receded, Earth's gravitational pull would decrease as the square of the distance. Also as it receded, the fraction of the total distance represented by the diameter of the Moon would decrease in proportion to the distance. The tidal effect would decrease for both reasons, and if both are taken into account it means that the tidal effect falls off as the *cube* of the distance.

The Sun is twenty-seven million times as massive as the Moon. If both Sun and Moon were at an equal distance from the Earth, the Sun's tidal effect upon the Earth would be twenty-seven million times that of the Moon's tidal effect upon the Earth.* The Sun, however, is 389 times as far from the Earth as the Moon is. The Sun's tidal effect is weakened by an amount equal to 389 x 389 x 389, or 58,860,000. Divide 27 million by 58,860,000 and we find that the Sun's tidal effect on Earth is only about 0.46 that of the Moon. If the Moon's tidal effect has not sufficed to slow the Earth's rotational period very much as yet, the Sun's certainly would not.

Mercury is closer to the Sun than the Earth is, and that would be a factor that would tend to increase the tidal effect of the Sun. On the other hand, Mercury is smaller than the Earth, and that would tend to decrease it. Taking both factors into account, it turns out that the Sun's tidal effect on Mercury is 3.77 times that of the Moon's tidal effect on the Earth, and only $\frac{1}{6}$ Earth's tidal effect on the Moon.

* This is a hypothetical case only, for if the centre of the Sun were as close to the Earth as the centre of the Moon is, the Earth would be far beneath the surface of the Sun.

The Sun, therefore, slows Mercury's rotation more effectively than the Moon slows Earth's, but less effectively than the Earth slows the Moon's. We might suspect, then, that Mercury rotates slowly but not so slowly as to face one side only to the Sun.

In 1890, Schiaparelli (who reported the canals on Mars thirteen years before) undertook the task of observing Mercury's surface. This is a very difficult thing to do, since Mercury is farther from us than Mars, usually; since Mercury shows only a crescent phase, usually, whereas Mars is always full or nearly full; and since Mercury, unlike Mars, is usually close enough to the brightness of the Sun to make comfortable viewing unlikely. Nevertheless, from what faint spots Schiaparelli could make out on the surface of Mercury, he decided that it rotated only once in each revolution of eighty-eight days, and that it faced only one side to the Sun.

In 1965, however, radar waves that were emitted from Earth were bounced off Mercury's surface. The echo, received on Earth, told a different story. The length of the radar waves changes if they strike a rotating body, and the change varies with the speed of rotation. From the nature of the reflected radar waves, it turns out that Mercury's period of rotation is fifty-nine days, or just $\frac{2}{3}$ of its period of revolution. This is a comparatively stable situation, not as stable as having its rotation equal to its period of revolution, but stable enough to resist further change through the Sun's insufficiently strong tidal effect.

Now we can return to the imaginary situation of our midget star, with Earth circling it at a distance of 300,000 kilometres (186,000 miles) from its centre. This distance is only $\frac{1}{500}$ that of our Earth from the Sun, and even allowing for the fact that the midget star had only $\frac{1}{16}$ the mass of the Sun, its tidal effect on Earth would be 150,000 times that of the Earth's tidal effect on the Moon.

There is no question, then, but that if Earth were close enough to a midget star to be within its ecosphere, the powerful tidal effect of the star would slow its rotation, and quite early in its lifetime cause it to face one side forever towards the star and one side forever away.

On the side facing always towards the star, the temperature would go up past the boiling point of water. On the side facing always away from the star, the temperature would drop far below the freezing point of water. There would be no liquid water on either side.

One could imagine that there might be a 'twilight zone' on the boundary between the forever-lit and the forever-dark hemispheres, in which the conditions would be mild. This would be so only if the orbit of the planet were nearly circular. Even then, the temperature on the hot side might be hot enough to result in the slow loss of the atmosphere, so that the planet would be airless and the twilight zone no more habitable than any other part.

As we imagine a larger and larger star, the ecosphere would be farther and farther from it. A planet within the ecosphere would be subjected to a smaller and smaller tidal effect. Eventually, if the star were large enough, the tidal effect will no longer be large enough to render the planet unfit for life as we know it.

We might estimate that a star should have at least $\frac{1}{3}$ the mass of the Sun (which means it would have to be of spectral class M2 at least) before a planet in its ecosphere would be suitable for life.

Nor is the matter of tidal effect the only problem with midget stars. The width of an ecosphere depends on how much energy a star is radiating. A massive, luminous star has an ecosphere far out in space and one that is very deep; deeper than the entire width of our Solar System. A midget star has an ecosphere that is close in on itself and is very shallow. The chance of a planet's happening to form within so shallow an ecosphere is vanishingly small.

Finally, stars smaller than spectral class M2 are very often 'flare stars'. That is, flares of unusually bright and hot gas periodically burst out on its surface. This happens on all stars, even on our Sun, for instance. On the Sun, however, such a flare would only add a small and bearable fraction to the ordinary Solar output of light and heat. The same flare on a dim midget star would increase its light and heat output by up to fifty per cent. A planet receiving a proper amount of energy from the midget star would receive far too much under flare conditions. The star's role as incubator would be carried out in too irregular a fashion to be compatible with life.

Between tidal effects, shallowness of ecosphere, and periodic flares, the exclusion of midget stars from further consideration in connection with extraterrestrial intelligence is triply justified.

Just right

If the stars with too much mass to serve as adequate incubators for life, those more massive than spectral class F2, make up a small fraction of all the stars, this is not the case for the stars that are less massive than spectral M2 and also don't serve as adequate incubators for life. Midget stars are very common. More than two-thirds of the stars in our Galaxy, and presumably in any Galaxy, are too small to be suitable for life.

Between spectral classes F2 and M2 are the stars that range in mass from 1.4 times that of the Sun to 0.33 times that of the Sun. At the upper end of this range, the lifetime of the stars is barely enough to give intelligence a fair chance to evolve. At the lower end of this range, a planet barely escapes tidal effects of too serious a nature.

Within the range, though, are the 'Sunlike stars', which, all other things being equal, are suitable incubators for life. While these Sunlike stars do not make up a majority of the stars in the sky, they are not really few in number, either. Perhaps twenty-five per cent of all the stars in the Galaxy are sufficiently Sunlike in character to serve as adequate incubators of life.

That gives us our third figure:

3 The number of planetary systems in our Galaxy that circle Sunlike stars = 75,000,000,000.

8 Earthlike planets

Binary stars

A star may be Sunlike and yet still not be a suitable incubator for life. It may have properties, other than its mass and luminosity, that make it impossible for an Earthlike planet to circle it.

A star may be like the Sun in every apparent respect, for instance, and yet have as a companion not a planet or a group of planets, but another star. The presence of two stars in close association may conceivably ruin the chances for an Earthlike planet to circle either one.

The possibility of multiple stars did not dawn on astronomers until about two centuries ago. After all, our Sun is a star without stellar companions and that made it seem a natural condition. When the stars were recognized to be other suns, they, too, were assumed to be single. To be sure, there are stars that are close together in the sky. For instance, Mizar, the middle star in the handle of the Big Dipper, has a fainter star, Alcor, very near it. Such 'double stars' were taken, however, to be single stars lying nearly in the same direction from the Earth but at radically different distances. In the case of Mizar and Alcor, this turned out to be true.

In the 1780s, William Herschel began to make a systematic study of double stars in the hope that the brighter (and presumably closer) one might move slightly and systematically with reference to the dimmer (and presumably more distant) one. This motion might reflect the motion of the Earth about the Sun and be the star's 'parallax'. From this, the star's distance could be determined, something that had not yet been done.

Herschel did find motions among such stars, but never of

the kind that would indicate the presence of a parallax. Instead, he found some double stars to be circling about a natural centre of gravity. These were true double stars, bound to each other gravitationally, and were called binary stars, from a Latin word meaning *in pairs*.

By 1802, Herschel was able to announce the existence of many such binary stars, and they are now known to be very common among the stars of the Universe. Among the bright and familiar stars, for instance Sirius, Capella, Procyon, Castor, Spica, Antares, and Alpha Centauri are all binaries.

In fact, more than two stars might be held together gravitationally. Thus, the Alpha Centauri binary (which are referred to as Alpha Centauri A and Alpha Centauri B) have a very distant companion, Alpha Centauri C, some 1,600,000,000,000 kilometres (one trillion miles) from the centre of gravity of the two other stars. A binary star system may also be gravitationally bound to another binary star system, the two pairs of stars circling a common centre of gravity. Systems of five or even six stars are known.

In every case, though, where more than two stars are involved in a multiple system, the stars exist in relatively close pairs widely separated from companion singles or other binaries.

In other words, suppose that there were a planet about Star A, which is a member of a binary system. Star B might be close enough to have some important effect on the planet. It might add its own radiation to that of Star A in different amounts at different times. Or else its gravitational pull might introduce irregularities into the planet's orbit that might not have existed otherwise.

On the other hand, if the A-B binary had, associated with it, a third star, or another binary, or both a star and a binary – all would be so far off that they would simply be stars in the sky without particular influence on the development of life on the planet.

From the standpoint of this book, therefore, let us talk only of binaries.

There is nothing puzzling about the existence of binaries.

When an initial nebula condenses to form a planetary system, one of the planets may, by the chance of the turbulence, attract enough mass to become a star itself. If, in the course of the development of our own Solar System, Jupiter had accumulated perhaps sixty-five times as much mass as it did, the loss of that mass to the Sun would not have been particularly significant. The Sun would still have very much the appearance it now has, while Jupiter would be a dim 'red dwarf' star. The Sun would then be part of a binary system.

It is even quite possible that the original nebula might condense more or less equally about two centres to form stars of roughly equal mass, each smaller than our Sun, as in the case of the 61 Cygni binary system; or each roughly equal in size to our Sun, as in the case of the Alpha Centauri binary system; or each larger than the Sun, as in the Capella binary system.

The two stars might, if they are of different mass, have radically different histories. The more massive star may leave the main sequence, expand to a red giant, and then explode. Its remnants would then condense to a small, dense star, while the less massive companion star remains on the main sequence. Thus, Sirius has as a companion a white dwarf, a small, dense remnant of a star that once exploded. Procyon also has a white dwarf as a companion.

The total number of binaries in the Galaxy (and presumably in the Universe generally) is surprisingly large. Over the nearly two centuries since their discovery, the estimate of their frequency has steadily risen. At the moment, judging from the examples of those stars close enough to ourselves to be examined in detail, it would seem that anywhere from 50 to 70 per cent of all stars are members of a binary system. In order to arrive at a particular figure, let us take an average and say that 60 per cent of all stars and, therefore, of all Sunlike stars, too, are members of a binary system.

If we assume that any Sunlike star can form a binary with a star of any mass, then, keeping in mind the proportions of stars of various masses, we could venture a reasonable division of the seventy-five billion Sunlike stars in the Galaxy as follows:

30 billion (40 per cent) are single

25 billion (33 per cent) form a binary with a midget star

18 billion (24 per cent) form binaries with each other

2 billion (3 per cent) form a binary with a giant star

Ought we now to eliminate the forty-five billion Sunlike stars involved in binary systems as unfit incubators for life?

Certainly, it would seem that we can omit the two billion Sunlike stars that form binaries with giant stars. In their case, long before the Sunlike star has reached an age where intelligence might develop on some planet circling it, the companion star would explode as a supernova. The heat and radiation of a nearby supernova is quite likely to destroy any life on the planet that already existed.

What about the remaining forty-three billion Sunlike stars forming a part of binaries?

In the first place, can a binary system possess planets at all?

We might argue that if a nebula condenses into two stars, the two will be twice as effective in picking up debris as one would be. Any planetary material that might escape one would be picked up by the other. In the end, therefore, there would be two stars and no planets.

That this is not necessarily so is demonstrated by the star 61 Cygni, the first whose distance from Earth was determined, in 1838, and that is now known to be 11.1 light-years from us.

61 Cygni, as I have said earlier, is a binary star. The two component stars, 61 Cygni A and 61 Cygni B, are separated by twenty-nine seconds of arc as viewed from Earth (a separation about $\frac{1}{64}$ the width of the full moon).

Each of the component stars is smaller than the Sun, but each is large enough to be Sunlike. 61 Cygni A has about 0.6 times the mass of the Sun, and 61 Cygni B about 0.5 times the mass. The former has a diameter of about 950,000 kilometres

(600,000 miles) and the latter a diameter cf about 900,000 kilometres (560,000 miles). They are separated by an average distance of about 12,400,000,000 kilometres (7,700,000,000 miles), or a little over twice the average distance between the Sun and Pluto, and they circle each other about their centre of gravity once in 720 years.

If we imagined the planet Earth circling one of the 61 Cygni stars at the same distance it now circles the Sun, the other 61 Cygni star would appear in the night sky at various times as a bright, starlike object, showing no visible disc, delivering no significant amount of radiation, and producing no significantly interfering gravitational effect.

Indeed, we might easily imagine each 61 Cygni star as possessing a planetary system nearly as extensive as the Sun's, each without interference from the other.*

In this particular case, we need not resort entirely to speculation. The very first planetary object about another star for which some evidence was obtained involved 61 Cygni. From the manner in which the separation of the two stars changed in a wobbly manner as they circled each other, the presence of a third body, 61 Cygni C, was deduced. From the extent of the wobble, it was thought to be a large planet some eight times the mass of Jupiter.

Soviet astronomers at the Pulkovo Observatory near Leningrad have studied the orbits of the 61 Cygni stars with care, have measured the irregularities of the wobble itself, and have suggested, in 1977, that *three* planets are involved. They feel that 61 Cygni A has two large planets, one with six times the mass of Jupiter and one with twelve times the mass, while 61 Cygni B has one large planet with seven times the mass of Jupiter.

These are very borderline observations. The tiny changes

* To be sure, if the Earth were as far from either 61 Cygni star as it is from the Sun, Earth would be frozen into a permanent ice age. On the other hand, if it were imagined to be at the distance from either star that Venus is from the Sun, Earth might do very well.

in the motion of the 61 Cygni stars can just barely be made out, and the chance that insignificant errors of measurement or interpretation have produced them is all too likely.

For what it's worth, however, and until something better comes along, it implies that both stars of a binary system (both stars being Sunlike stars) have planets – large planets at least. If large planets exist, however, it doesn't take much of a strain to suppose the existence of a large collection of smaller planets, satellites, asteroids, and comets – all too small to leave detectable marks on the wobble.

Of course, some binary systems are separated by smaller distances than the 61 Cygni stars.

Consider the two stars of the Alpha Centauri binary system. Alpha Centauri A has a mass 1.08 times that of the Sun, and Alpha Centauri B a mass 0.87 times that of the Sun. The two stars are separated by an average distance of 3,500,000,000 kilometres (2,200,000,000 miles). They revolve about the centre of gravity in quite elliptical orbits, however, and are much closer to each other at some times than at others. The maximum distance between the two stars is 5,300,000,000 kilometres (3,400,000,000 miles) and the minimum distance between the two is 1,700,000,000 kilometres (1,050,000,000 miles).

Suppose we imagined Alpha Centauri B circling our Sun exactly as it, in fact, circles Alpha Centauri A. If we plotted Alpha Centauri B's orbit relative to the Sun, it would follow an elliptical path that would carry it well beyond the orbit of Neptune at its farthest recession from the Sun, and nearly as close as the orbit of Saturn at its nearest approach.

Under such circumstances, neither star could have a very extensive planetary system of the sort the Sun has now. Planets at the distance of Jupiter or the other giants, circling either star, would be interfered with by the gravitational influence of the other star and would have unstable orbits.

On the other hand, an inner planetary system might still exist. If Alpha Centauri B were circling our Sun as it circles Alpha Centauri A, we on Earth could scarcely tell the difference with our eyes closed. Alpha Centauri B would be a bright, starlike object in the sky, which at its closest approach would be 5,000 times brighter than our full Moon and $\frac{1}{100}$ as bright as our Sun. It would add anywhere from 0.1 per cent to 1 per cent to the heat we receive from the Sun, depending on what part of its orbit it was in, and we could live with that. Nor would its gravitational influence affect Earth's orbit in any significant way.

For that matter, Alpha Centauri B could have an inner planetary system, too. A planet circling in its ecosphere (which would of course be closer to itself than the ecosphere is to either Alpha Centauri A or the Sun) would not be seriously interfered with by its somewhat larger companion.

As in the case of the 61 Cygni system, both Alpha Centauri A and Alpha Centauri B would have what we might call a 'useful ecosphere', one in which an Earthlike planet could orbit without serious interference from the companion in terms of either radiation or gravitation.

Robert S. Harrington of the US Naval Observatory in 1978 reported the results of high-speed computer studies of orbits about binary stars.

If a Sunlike star is part of a binary system, and if the separation between the two stars is at least 3.5 times the distance of the ecosphere from the Sunlike star, then it is a useful ecosphere. In the case of our own Solar System, it would mean that the Sun could have a companion at a distance equal to that of the planet Jupiter, without interfering with Earth gravitationally. If the companion were somewhat less luminous than Alpha Centauri B, it would not interfere with Earth significantly as far as radiation was concerned.

There are binary systems with stars even closer together than those of the Alpha Centauri system. The two stars of the Capella binary system are separated by a distance of only 84 million kilometres (52 million miles) or less than the distance of Venus from the Sun.

Neither star in such a binary could have a planetary system in the Sun's sense. Planetary orbits about one of the stars would be interfered with gravitationally by the other and the orbit would not be stable.

If a planet were far enough away, however, it would not circle either one star or the other but would circle, instead, about the centre of gravity of the two stars. Such a planet would treat the two stars gravitationally as a single dumbbell-shaped object.

Harrington calculates that a planet whose distance from the centre of gravity of the binary system was equal to at least 3.5 times the distance of separation between the two stars would have a stable orbit. In the case of the Capella system, a planet to have a stable orbit would have to be at least 300 million kilometres (185 million miles) from the centre of gravity.

In a close binary system, where the two stars are of the proper total luminosity, such an outer orbit might well be within the ecosphere of the two stars taken together. This is another way in which a binary might have a useful ecosphere.

There are pairs of stars that circle each other so closely that our best telescopes cannot make them out as separate stars. Their existence as pairs is given away by the spectroscope, when the dark lines of the spectrum sometimes double, rejoin, double, rejoin, and so on, over and over.

The simplest explanation is to suppose that there are two stars very close together and circling each other, so that one is receding from us while the other is approaching us. In that case, one would produce a red shift, while the other was simultaneously producing a violet shift, and that is why the lines would appear to double. It is the same principle that causes the lines of a rotating star to broaden. The revolution of two stars is more rapid than the rotation of one star, so that in the latter case the broadening is carried on to the point of actual spreading apart into two lines.

The first such 'spectroscopic binary' to be discovered was Mizar, and it was in 1889 that the American astronomer Edward Charles Pickering (1846–1919) detected the doubling of its spectral lines. Actually, the component stars of Mizar are separated by 164 million kilometres (102 million miles), which is a larger separation than that of the stars of the Capella system. The Mizar pair fail to be seen as a pair in the telescope because the system is so far away.

The component stars of some spectroscopic binaries are much closer to each other than that. They can be within a million kilometres of each other, almost touching, and making a complete circle about the centre of gravity in a couple of hours.

If we could imagine the Sun replaced by two stars, each half as luminous as the Sun and separated by less than 42,700,000 kilometres (26,500,000 miles) – somewhat less than the distance between the Sun and Mercury – the Earth would remain stably in its orbit. Planets at the distance of Mercury and Venus could not, under those conditions, remain in stable orbit, but Earth could.

In such a case, of course, the sum of the mass of the two stars would be greater than that of the Sun, and Earth's period of revolution would be considerably less than a year. In addition, with two separate stars at changing distances, Earth's seasons would show more complicated variations, perhaps, than they now do. Neither of these two factors, however, need render Earth unsuitable for life.

Well, then, how many Sunlike stars in our Galaxy have useful ecospheres?

To begin with, we may fairly assume that all the Sunlike stars that are single have useful ecospheres, and that means thirty billion right there.

Of the binary systems we have eliminated all Sunlike stars that have as a companion a giant star (or a small, dense star that is the shrunken and condensed remnant of a giant star that exploded).

Of the eighteen billion Sunlike stars that are in binary association with another Sunlike star, we might estimate conservatively that only one-third have useful ecospheres. That would mean six billion stars in this category. At a guess (nothing more than that) I would say there would be four billion binaries with two Sunlike stars, in which only the larger would have a useful ecosphere; and one million binaries of this kind in which both Sunlike stars would have a useful ecosphere.

Finally, what of the binaries in which a Sunlike star is teamed with a midget star? We had estimated there were twenty-five billion such binaries in the Galaxy altogether. A midget star is far less likely to interfere with a planetary system, either gravitationally or radiationally, than a larger star would. We might estimate, again conservatively, that two-thirds of these Sunlike stars have useful ecospheres, and this would mean approximately sixteen billion stars.

We now have our fourth figure:

4 The number of Sunlike stars in our Galaxy with a useful ecosphere = 52,000,000,000.

Star populations

Yet we are not through. A Sunlike star may have a useful ecosphere and even so there may be no possibility of an Earthlike planet revolving within that ecosphere. As it happens, stars may differ in ways other than mass, luminosity, and state of association. They may also differ in chemical composition.

When the Universe first formed about fifteen billion years ago, matter seems to have spread outwards from an exploding central mass. To begin with, that matter consisted almost entirely of hydrogen, the simplest element, with a small admixture of a few per cent of helium, the next simplest element. Virtually none of the still heavier elements existed.

This primordial matter, forming a Universe-sized mass of gas, split up into turbulent sections, each of galaxy size. Out of these protogalaxies, the stars of the various galaxies formed.

If we concentrate on any of the galaxy-sized masses of gas, the central regions were denser than the outer regions. The gas in the central regions split up into small, star-sized masses pretty evenly, each crowding the other so that no one star-sized mass had more chance than another to collect its share.

The result was that very many stars were formed, all small and medium in size; virtually none of them giants. What's more, nearly all the gas was collected by one star or another, so that the interstellar regions in a galactic centre ended up almost gas free.

These stars, characteristic of the central regions of a galaxy, are called Population II stars.

For regions at moderate distance outside the centre, there is not enough gas to form a steady, continuous packing of stars. The gas shreds into a couple of hundred smaller pockets of denseness, however, and out of each of them a tight group of some ten thousand to a million stars form. In this way, a 'globular cluster' is formed. Globular clusters are arranged in a spherical shell about the galactic centre, and are virtually dust free; the stars in such clusters are also Population II in nature.

The point to remember about Population II stars is that they were formed out of a gas that was largely hydrogen, with a little bit of helium, and virtually nothing else. The planetary systems that formed about such stars must be made up of planets that are also of that chemical structure. What planets do form about Population II stars would rather resemble Jupiter and Saturn in composition, but would lack the ad-mixture of ices – water, ammonia, methane, and so on – that those planets possess.

There would be no small objects in the planetary systems, since small objects would not have enough gravitational pull to retain the hydrogen and helium which were alone available.

Nor would there be life, for to have life (as we know it) we need such elements as carbon, oxygen, nitrogen, and sulphur, which are not present in appreciable amounts in Population II planetary systems.

Of course, the heavier elements do form with time. As each Population II star burns over the course of billions of years, heavy elements build up in its core through fusion reactions, including particularly those needed for life.

These heavier elements are, however, useless for the production of life as long as they remain at the core of stars.

Eventually a star leaves the main sequence, however, expands, and then collapses. If the star is a small one and not too much larger than our Sun, the process of collapse is not accompanied by an explosion, and a white dwarf is produced. In the process of collapse, however, up to one-fifth of the mass of the collapsing star is left behind as a cloud of gas surrounding the white dwarf. The result is what is called a planetary nebula. The expanding shell of gas slowly spreads through space until it becomes too rarefied to detect visually, and left behind is a bare white dwarf.

If a star is more massive than 1.4 times the mass of the Sun, it explodes as it collapses. The more massive the star, the more violent the explosion. Such a supernova explosion can eject up to nine-tenths of the mass of a star into space as swirls of gas.

The gas spreading into space, whether it started as the product of a planetary nebula or of a supernova, contains appreciable percentages of the more complicated elements. The process of supernoval explosion would, in particular, manufacture the really complex elements, which do not form in the centre of stars that are quietly maturing on the main sequence. In the centre of those stars, nothing past iron is produced, whereas in the comparatively brief episode of the supernova explosion, elements up to uranium and beyond are produced.

The Population II stars, however, are not very massive and, containing as they do a high percentage of hydrogen to begin with, they remain on the main sequence for a long time. Even in the fifteen billion years that have elapsed since the big bang, almost all those stars are still on the main sequence and the heavy elements remain tucked inside their cores.

From all this we might deduce that the centres of galaxies are quiet, uneventful places – and we would be wrong.

In 1963, quasars were discovered. These are starlike objects; indeed, when first discovered they were thought to be dim stars of our own Galaxy. They turned out, instead, to be located at distances of over a billion light-years, farther than any of the visible galaxies. To be visible at that distance, quasars had to be shining with the luminosity of 100 ordinary galaxies. Yet they are small objects, at most one or two light-years across, as compared with the diameters of many thousands of light-years that characterize ordinary galaxies.

The evidence now seems to favour the thought that quasars are bright galactic centres, surrounded, of course, by the outer structure of an ordinary galaxy. At the huge distance of the quasars, however, only the bright centre is visible.

The question, then, is: What makes a galactic centre blaze so brightly?

It would appear that the very centres of galaxies are quite commonly the sites of violent events. Some are visibly exploding; some giving off vast streams of radio waves from sources on either side of the centre as though an explosion has ejected material in opposite directions.

All galactic centres are bright; some are brighter than others. As we consider galaxies that are more and more distant, we reach a point where we see only the brightest of the bright galactic centres – the quasars.

What happens to the quiet Population II stars to initiate such violence?

If they were left to themselves, nothing; but they are not left to themselves. In the crowded precincts of the galactic centres, the stars are a million times as densely packed as in our own area of the galactic outskirts. The stars at the galactic centre may be separated by average distances of only 70 billion kilometres (45 billion miles), only ten times the distance between the Sun and Pluto.

Under such packed conditions, collisions and near-collisions may not be very rare. Transfer and capture of mass may serve to build up stars of great mass that quickly explode with a force that leads to a veritable chain reaction of explosions and to the formation of 'black holes'. These are the ultimate in star condensations (see my book, *The Collapsing Universe*).

A black hole is matter at its ultimate density, and has a gravitational field so intense at its surface that nothing can escape it, not even light.

If a black hole is formed under conditions in which matter of all kinds surrounds it (as in galactic centres), such matter is constantly spiralling into the black hole, releasing x-rays and other energetic radiation in the process. (This radiation is released before the matter actually enters the black hole, so that it can escape into outer space.) The black hole gains in mass and may eventually be large enough to swallow stars whole.

There is a strong radiation source at the very centre of our own Galaxy, and it may well be that a black hole is present there, one that has a mass of 100 million stars. The giant galaxy M87 was reported in 1978 to have a black hole in its centre in all likelihood, one that has a mass as high as that of ten billion stars. It may even be that every galaxy and every globular cluster has a black hole at its core.

Such violent events at the centres of galaxies may produce the massive atoms of complex elements and spread them through space, but of what use would that be? Those violent events are the sites of emission of enormous quantities of energetic radiation, and for many light-years in every direction, life (as we know it) might for that reason be impossible.

The Population II regions are therefore, considering chemical constitution or energetic radiation, doubly unsuitable for life.

Suppose we pass on to the outskirts now, regions where the violence and radiation of the centre does not reach.

Here, the primordial gas was relatively thin and was distributed irregularly. For that reason, stars were formed irregularly, and giant stars were routinely formed in numbers that could not possibly have existed in the centre. (Of course, many medium and small stars were also formed.)

The stars in the outskirts of a galaxy, rich in giants and spread out irregularly over much vaster volumes of space than exist in the central regions, are referred to as Population I stars.* What's more, there were places in the outskirts where

* It is because the stars of our own region of the Galaxy are of this type that they got the 'I' classification.

the gas was too thin to condense readily. To this day, therefore, the outer Population I regions of the galaxies are rich in clouds of gas and dust.

The original Population I stars were as entirely hydrogen-helium in constitution as were the Population II stars. There was this difference, however.

The giant stars that formed in the galactic outskirts didn't remain on the main sequence long. A few hundred thousand years only, for the real monsters; a few million years for the mere titans; as much as a billion years for those that were simply giant.

And when they left the main sequence, expanded and finally collapsed, they exploded into supernovas of unimaginable violence. Vast volumes of gas, containing significant quantities of complex elements rolled out into space, adding themselves to the clouds of uncondensed gas that were already present.

Such explosions take place repeatedly in the outer regions of a galaxy, but so widely separated are the stars in those vast outer regions that supernovas do not seriously affect any stars other than (at most) their immediate neighbours.

As many as 500 million supernova explosions may have taken place in the outskirts of our own Galaxy since it came into being. The 500 million have enriched space enormously with complex elements, and have added to the density of the clouds of gas and dust that existed from the beginning. The outward force of the explosion may even have served as an initiation of swirls and compressions in nearby gas clouds that led to the formation of a new star, or whole groups of new stars.

New stars, forming out of gas clouds containing elements produced in an older star that had distributed those elements in its death throes, are called second-generation stars. Our Sun, which formed only five billion years ago, when the Galaxy was already ten billion years old and after hundreds of millions of stars had already died, is a second-generation star.

The cloud out of which second-generation stars are formed contain the elements out of which ices, rocks, and metals are formed, and therefore can produce planetary systems similar to our own Solar System.

If we look for Sunlike stars that are capable of incubating life, therefore, we must eliminate Population II stars and even many of the Population I stars. We can only consider second-generation Population I stars.

Population II stars are confined to only a small portion of the total volume of a galaxy, to its compact central regions and to the almost as compact globular clusters. All the open vastness of the outer regions is the domain of Population I stars.

That is not, however, as impressive as it sounds. Some eighty per cent of the stars of a galaxy are to be found in the compact central regions and in the globular clusters.

We might argue, too, that only half of the twenty per cent of stars that are in the Population I regions are second-generation stars. That means that only ten per cent of all the Sunlike stars with effective ecospheres are second-generation Population I stars, and can conceivably have Earthlike planets revolving about them.

That gives us our fifth number:

5 The number of second-generation, Population I, Sunlike stars in our Galaxy with a useful ecosphere = 5,200,000,000.

The ecosphere

Even if a star is a perfect incubator, if it is the precise duplicate of our Sun in every respect, that is still not enough. What is needed is not only an incubator, but something to be incubated as well. In short, there must be a planet on which life can develop in the beneficent radiation of the star it circles.

To be sure, we have already decided that virtually every star has its planetary system, so that there are 5,200,000,000 second-generation, Population I, Sunlike stars in our Galaxy with planets – but where are those planets located?

A given star might be a perfect incubator, but some of its planets may be too close to it and therefore too hot to bear life, while others might be too far and therefore too cold to bear life. There might be no planet at all within the star's ecosphere on which water could exist as a liquid.

What are the chances, then, that a given star has a planet, at least one, within its ecosphere?

In trying to make a judgement here, we are badly hampered by the fact that we know only one planetary system in detail – our own. What's more, we have no way at all at present of possibly learning any appropriate details about any other planetary system. The few planets we may possibly have detected circling nearby stars are all the size of Jupiter or larger.

Such giant planets are the only ones we can possibly detect at the moment, and that only with great difficulty and considerable uncertainty. Whether there are any planets actually within the ecosphere of such stars, planets that lie closer to the star and that are small enough to be Earthlike, it is impossible to tell.

We are forced to fall back on the only thing we have, our own planetary system. It may conceivably be a very atypical, freakish structure that simply can't be used as a guide, but we have no reason to think so. The temptation is to follow the principle of mediocrity and to suppose that the planetary system in which we find ourselves is a typical one and that it can be used as a guide.

There is some hope that this is not just prejudice on our part, or wishful thinking. The American astronomer Stephen H. Dole has checked this, as well as one can, by computer. Beginning with a cloud of dust and gas of the mass and density thought to have served as the origin of the Solar System, he set up the requirements for random motion, for coalescence on collision, for gravitational effects, and so on. The computer calculated the results.

The computer worked out different random happenings, and in every case a planetary system very much like ours resulted. There were from seven to fourteen planets, with small planets near the Sun, large planets farther out, and small planets again still farther out. In almost every case, there was a planet rather like the Earth in mass, at rather like Earth's distance from the Sun, and planets much like Jupiter in mass at much like Jupiter's distance from the Sun, and so on.

In fact, if a diagram of the real Solar System is mixed in with the various computer simulations, it is not at all easy to separate the real from the simulated.

It is hard to say how much importance we can lend to such computer simulations, but for what they are worth, they do give a colour of truth to the principle of mediocrity, at least in this respect.

If we now study our own planetary system on the assumption that it is typical, we can see that the planets move in nearly circular obits that are widely spaced, and that the orbit of one does not overlap the orbit of the planet within or the one without.

This tends to make sense, since orbits that are too closely spaced would, in the long run, prove unstable. Between collisions and gravitational interactions, the worlds are bound to nudge themselves apart early in the history of the planetary system.

This means that it is completely unlikely that there will be very many worlds crammed into the ecosphere of a Sunlike star. The ecosphere is not likely to be wide enough for that. In fact, we might suspect intuitively that once the planets are done nudging themselves apart, not more than one planet is likely to find itself within the ecosphere; or two, if we find ourselves dealing with a double planet on the order of the Earth and the Moon.

How does this check with our own planetary system?

Here, for instance, Earth is clearly within the Sun's ecosphere, or you and I would not exist to question the matter.

Even as late as a generation ago, the ecosphere would have seemed to be some 100 million kilometres (62 million miles) deep at least, since it was generally supposed that while Venus might be uncomfortably warm and Mars uncomfortably cool, both had environments not so extreme as to preclude the presence of life.

Not so. Venus has suffered a runaway greenhouse effect and

is far too hot for life. Mars may be in a permanent ice age and be far too cold for life. The trigger leading in either direction may be a minor one.

If this is so, the Sun's ecosphere may be shallower than we think. Indeed, in 1978, Michael Hart of NASA simulated Earth's past history by computer and if his starting assumptions are correct, and his computer programming likewise, then it would seem that Earth, at one stage in its history, escaped a runaway greenhouse effect by a narrow margin and at another stage escaped a runaway ice age by a narrow margin. A little nearer the Sun or a little farther from it, and Earth would have fallen prey to one or the other. It may be, from Hart's figures, that the Sun's ecosphere is only ten million kilometres (6,200,000 miles) thick and it is only a most fortunate coincidence that Earth happens to be in it.

Well, then, what can we say? If the ecosphere is quite wide (even if not wide enough to include either Venus or Mars), then from Dole's computer simulation of planetary systems, a planet is virtually certain to form within it somewhere. The probability would be roughly 1.0.

On the other hand, if Hart's computer simulation of Earth's past history is accurate, then it is very likely that no planet at all will form within the ecosphere, and that all the planets near the star will be Venuslike or Marslike, and only on quite rare occasions Earthlike. The probability of a planet within the ecosphere would then be close to 0.0.

The results of computer simulation are still too recent and, perhaps, too crude to allow us to lean too certainly in either the optimistic or the pessimistic direction. It might be best to split the difference and to suppose that the probability of a planet within the ecosphere is close to 0.5, or 1 in 2.

That would give us our sixth number:

6 The number of second-generation Population I stars in our Galaxy with a useful ecosphere and a planet circling it within that ecosphere = 2,600,000,000.

Habitability

The mere fact that a planet is in the ecosphere does not mean that it is a suitable abode for life; that it is habitable, in other words.

For proof of that we need look no farther than our own Solar System. The Earth itself is the only planet in the Solar System that is clearly within the ecosphere of the star it circles. Our definition of the word *planet*, however, obscures the fact that there are *two* worlds in the ecosphere just the same.

The Moon, strictly speaking, is not a planet, because it circles the Earth (or rather the Earth-Moon centre of gravity, which the Earth also circles), but it is a world. What's more, it is a world that is just as firmly within the ecosphere as the Earth and yet the Moon is not a habitable world.*

The Moon clearly has too little mass to be habitable, since because of its small mass it cannot retain an atmosphere or liquid water. What, then, can we say about the masses of planets?

As I have said in the case of Population II stars where the only materials for planetary structure are hydrogen and helium, the only possible planets would seem to be giants with the mass of Uranus or more. Nothing less would possess the gravitational intensity that would make it possible to hold on to hydrogen and helium.

In the case of Population I stars, which are the only ones we are considering as suitable incubators for life, we have metals, rocks, and ices in addition to hydrogen and helium for uses as structural materials. Again here, only giant planets can make use of the hydrogen and helium, and it is precisely because they can that they are giant planets.

* We judge the habitability of a world by the fact that life can originate on it and be maintained on it independently of other worlds. If human beings eventually establish a base on the Moon, that should be credited not to the Moon's habitability but to human ingenuity and technology.

On the other hand, where Population I stars are concerned, smaller worlds of all sizes can be built up of metals, rocks, and ices, since these will hold together through forces other than gravitational.

How large can these smaller worlds be?

Not very large, for even among Population I stars of the second generation, the quantity of materials other than hydrogen and helium is rather small, and cannot be used to build a large world. And if these stars could, they would gather hydrogen and helium and become giant worlds.

Dole's computer simulations of planetary formation make it seem pretty clear that within the ecosphere of Sunlike stars those planets that are not giants are quite small.

How large and massive can a nongiant planet be?

If we exclude the four giant planets of the Solar System (and the Sun itself, of course), then the largest body in the Solar System is none other than the Earth itself.

Earth is, therefore, very likely to be near the top limit of mass for nongiant, nonhydrogen planets.

A planet somewhat larger than Earth, but not much larger, would, if all other factors were suitable, surely be habitable. The one unavoidable consequence of the greater mass would be a more intense gravitational field, which might manifest itself as a somewhat higher surface gravity. There is no reason to think that life could not adapt itself to a somewhat higher surface gravity.

After all, life on Earth evolved in the ocean where, thanks to buoyancy, the influence of gravity is minor. Living organisms invaded the dry land, where the influence of gravity is major, yet not only coped with it but even evolved ways of moving rapidly despite gravity. A somewhat greater surface gravity would surely not defeat life when it has shown such amazing adaptability on the one world where we can study it in detail.

Then, too, if a world is somewhat more massive than Earth, but also somewhat less dense, so that its surface is farther from the centre than one would expect under Earthlike conditions, the surface gravity may be no higher than that of Earth, or even a bit lower.

We might reasonably conclude, then, that in the ecosphere where a star's heat will be great enough to preclude the gathering of hydrogen and helium, planets will not form that are too massive for life.

Worlds that are not massive enough can certainly form, as for instance the Moon, but how massive is not massive enough?

To support life, a world must be massive enough to generate a sufficiently large gravitational field to hold a substantial atmosphere – not so much for the sake of the atmosphere, as because that alone would make it possible to have free liquid on the surface.

In the Solar System there are exactly four of the nongiant worlds with substantial atmospheres: Earth, Venus, Mars, and Titan.

Venus, with a mass 0.82 that of the Earth, has a considerably denser atmosphere than Earth (but is nonhabitable for other reasons). Mars, which has 0.11 times the mass of the Earth, has a very thin atmosphere; one that, while substantial, is clearly not sufficient to support anything but, just possibly, the simplest forms of life. Titan, which has a mass 0.02 that of Earth, has an atmosphere that may be somewhat more substantial than that of Mars, but which exists at all only because Titan is far beyond the outermost reach of the ecosphere.

Within the ecosphere, a world can maintain an adequate atmosphere if it is not as massive as Earth, but it should certainly be more massive than Mars. If, let us say, its mass were 0.4 times that of Earth, that might be sufficient.

In or near the Sun's ecosphere, there are four worlds of considerable size: Earth, Venus, Mars, and the Moon. (There are also bodies of trifling size, such as the two satellites of Mars, and periodic entries of asteroids or comets, but these may all be ignored as not significant.) Of these four, Earth and Venus are higher in mass than the 0.4 mark, while Mars and the Moon are lower.

If we use the principle of mediocrity and consider this as a fair sample of the situation in the Universe as a whole, we could conclude that of all the worlds in or near appropriate ecospheres surrounding appropriate stars, only half have masses suitable for habitability.

If a world of the proper mass is present in the ecosphere, many of its characteristics would automatically be like those of the Earth. For instance, it would be too warm for substantial quantities of the icy materials to be in the solid state; and in liquid or gaseous state, the gravitational field of the world would not be intense enough to hold them. Therefore, a world of the proper mass in the ecosphere would be built up primarily of rock, or of rock and metal, as are all the worlds of the inner Solar System.

Water, as the icy material that melts and boils at the highest temperature, that is the most common, and that most readily combines with rocky substances, is on all three counts the most likely of the ices to be retained to some degree. Therefore, worlds of the proper mass in the ecosphere are very likely to have quantities of surface water in gaseous, liquid, and solid form. They would have oceans that would cover at least part of the surface.

In short, a world in the ecosphere that is of the proper mass would be 'Earthlike' in character.

If one out of every two worlds in the ecosphere is Earthlike, we have our seventh figure:

7 The number of second-generation, Population I, Sunlike stars in our Galaxy with a useful ecosphere and an Earthlike planet circling it within that ecosphere = 1,300,000,000.

Even an Earthlike planet, in terms of temperature and structure, might be nonhabitable for any of a variety of minor reasons. It could not very well support life if it were subjected, for instance, to great extremes of environmental conditions.

Suppose a planet had an average distance from the Sun that was right in the middle of the ecosphere, but suppose it also had a particularly eccentric orbit. At one end of its orbit it might swoop so far towards the Sun as to be well inside the inner border of the ecosphere, while on the other side it would recede so far from the Sun as to be well outside the outer border. Such a planet would have a short, unbelievably torrid summer that might briefly bring the oceans to a boil; and a long, unbelievably frigid winter, during which the oceans may begin to freeze.

We can imagine life might develop that could withstand such extremes, but it seems reasonable to suppose that the chances are it would not.

Again, extremes would lower the chances of life's coming into being if a planet's axis of rotation were inclined so steeply to the vertical (relative to its plane of revolution about its star) that the major portion of the planet would be in sunlight for half a year and in the dark for half a year.

And yet again, if a planet rotates very slowly, the days and nights are each long enough to allow undesirable temperature extremes.

If a planet is a little on the massive side, it may just happen to collect enough water to make its oceans a planetary one, with little or no dry land. Even if life then develops, it is not likely that technology will, and we are looking not for life alone, but technology as well.

In reverse, if a planet is a little on the nonmassive side and little water is collected, the world may be mostly desert, and life may at best form to only a limited extent and reach insufficient levels of complexity.

The atmosphere may not be quite right in some ways, and block off too much of the sunlight, or too little of the ultraviolet radiation. Or else the crust may not be quite right and there may be too much volcanic action or earthquakes. Or else the surroundings in near space may not be quite right and meteoric bombardments may be too intense for life to maintain itself.

None of these imperfections is very likely, perhaps. After all, among the planets of our Solar System, only two (Mercury and Pluto) have orbits that are significantly elliptical; only one (Uranus) has an enormous axial tilt; only two (Mercury and Venus) have very slow periods of rotation, and so on.

Yet although each one of the imperfections is unlikely in itself and may affect only one out of ten Earthlike planets, or fewer, all the various imperfections mount up.

Again, we might suppose (intuitively) that only one out of every two Earthlike planets is Earthlike in every important particular; that it has a day and night of reasonable length, seasons that do not go to unreasonable extremes, oceans that are neither too extensive nor too restricted, a crust that is neither too unsettled nor too geologically inert, and so on.

We might say that such planets are 'completely Earthlike' or, better, simply 'habitable'. In fact, we no longer have to specify that we are speaking of Sunlike stars, or of second-generation Population I stars, or of ecospheres. The term *habitable* would imply all that out of necessity.

If, then, one out of every two Earthlike planets are habitable, we have our eighth figure:

8 The number of habitable planets in our Galaxy = 650,000,000.

This sounds like a large number and, of course, it is, but it represents a measure of our conservatism also. This number means that in our Galaxy, only one star out of 460 can boast a habitable planet. What's more, it is a more conservative figure than some astronomers would suggest. Carl Sagan, who is one of the leading investigators of the possibility of extra-terrestrial intelligence, suggests there may be as many as one billion habitable planets in the Galaxy.

9 Life

Spontaneous generation

It is rather breathtaking to decide on the basis of (we hope) strict logic and the best evidence we can find that there are 650 million habitable planets in our Galaxy alone, and therefore over two billion billion in the Universe as a whole. And yet, from the standpoint of the subject matter of this book, of what value are habitable planets in themselves? If they lack life, their habitability comes to nothing.

Our calculations concerning extraterrestrial intelligence must therefore come to a halt right here, unless we can say something reasonable about the chance that a habitable planet actually has life on it.

In order to do that, we must again turn to something that is known, and that is the one habitable planet that we *know* to have life on it – Earth itself. In other words, before we can say anything sensible about life on habitable planets in general, we must be able to say something sensible about how life came to exist on the Earth.

Early speculations about the existence of life on Earth invariably assumed it to have been created through some non-natural agency, usually through the action of some god or demigod. The best-known story in our Western tradition is that humanity was created in the same series of divine acts that created the Universe generally.

In six days of creation the job was done. God created light on the first day; the land and sea on the second; plant life on the third; the heavenly bodies on the fourth; animal life of the sea and air on the fifth; and animal life on land on the sixth. As the last creative act on the sixth day, humanity was brought into being.

Life, created on three different days, was considered as having come into being in separate species ('after his kind' it says in the King James Bible). Presumably, these were the species that continued to exist into contemporary times. As some believed, no species were added to the first creation and none subtracted.

As to the date of this Divine creation, the Bible is not specific, for the habit of dating with compulsive precision is a rather late development in historical writing. Deductions based on various statements in the Bible, however, place the date of creation only a few thousand years in the past. The precise date usually found in the headings of the King James Bible is 4004 BC, this date having been worked out by the Irish theologian James Ussher (1581–1656).

Although the creation of the world (or of different worlds) was assumed to be a once-for-all act, it was common in early times to assume that this was not necessarily true for life.

Actually, this is a reasonable attitude. After all, while there was no visual evidence of any creation of worlds in the course of human history, there did seem to be visual evidence for the creation of living things without the intervention of earlier living things.

Field mice may make their nests in holes burrowed into stores of wheat, and these nests may be lined with scraps of scavenged wool. The farmer, coming across nests from which the mother mouse has had to flee, and finding only tiny, naked, blind infant mice, may come to the most natural conclusion in the world: he has interrupted a process in which mice were being formed from musty wheat and rotting wool.

Let meat decay and small wormlike maggots will appear in it. Frogs can seem to arise out of river mud.

If the notion were true for various species of vermin, it might be true for all species of organisms, though perhaps less common for the larger and more complex species such as horses, eagles, lions, and human beings.

In fact, if one were sufficiently daring, one might suppose that the tale in Genesis was a fable; that this sort of 'spontaneous generation' of living things from nonliving antecedents might account for the *original* beginning of life. Little by little each species might have formed, first the simple ones and later the more complex ones, with human beings, naturally enough, last of all.

And in that case, if we were to apply this to habitable planets generally, we would see that they, too, would naturally form life. All of them would bear life.

Provided, that is, the doctrine of spontaneous generation could withstand close examination – and it couldn't.

The first crack in the doctrine appeared in 1668, thanks to an Italian physician and poet named Francesco Redi (1626–1697). Redi noticed that decaying meat not only produced flies, but also attracted them. He wondered if there were a connection between the flies before and the flies after, and tested the matter.

He did this by allowing samples of meat to decay in small vessels. The wide openings of some vessels he left untouched; others he covered with gauze. Flies were attracted to all the samples, but could land only on the unprotected ones. Those samples of decaying meat on which flies landed produced maggots. The decaying meat behind the gauze, upon which the foot of fly never trod, produced no maggots at all, although it decayed just as rapidly and produced just as powerful a stench.

Redi's experiments showed plainly that maggots, and flies after them, arose out of eggs laid in decaying meat by an earlier generation of flies. There was no spontaneous generation of flies, just the normal process of birth from eggs (or seed).

Even as Redi was working out his demonstration, a Dutch biologist, Anton van Leeuwenhoek (1632–1723), was riding his hobby and grinding perfect little lenses (primitive microscopes, actually) through which he could look at tiny things and magnify them to easy visibility.

In 1675, he discovered living things in ditch water that were too small to be seen by the naked eye. These were the first 'microorganisms' known, and those that van Leeuwenhoek first discovered are now called protozoa, from Greek words meaning *first animals*. In 1680, van Leeuwenhoek discovered that yeast was made up of tiny living things even smaller than most protozoa, and in 1683 he observed still tinier living things, which we now call bacteria.

Where did these microscopic living things come from?

Broths were invented in which microorganisms could multiply. It turned out not to be necessary to seek microorganisms to place in these broths. A broth might be boiled and filtered until there was nothing in it that the lens of a microscope could detect. If one waited a while and looked again, the broth was inevitably swarming with life. (What's more, it was microorganisms that caused meat to decay even when no microorganisms were placed in the meat.)

Perhaps spontaneous generation did not take place in the case of those species visible to the unaided eye. In the case of the microorganisms – bits of life far simpler than the familiar plants and animals of everyday life – spontaneous generation might well be possible. In fact, it seemed established.

But then, in 1767, came the work of an Italian biologist, Lazzaro Spallanzani (1729–1799). He not only boiled broth, but he sealed off the neck of the flask containing it. The broth, boiled and sealed, never developed any form of microscopic life. Shortly after the seal was broken, however, life began to swarm.

A sealed neck, keeping out the air, acted like Redi's gauze, and the conclusions had to be similar to Redi's conclusions. There are microscopic and unseen creatures all about us in the air that are smaller and harder to observe than even the eggs of flies. These airborne bits of life fall into any broth left open to the air, and there they multiply. (Spallanzani isolated a single bacterium and watched it multiply by simply splitting in two.) If those bits of life are kept out of the broth, no life originates.

In 1836, a German biologist, Theodor Schwann (1810–1882), went even further. He showed that broth remained sterile even when open to air, provided the air to which it was exposed had been heated first in order to kill any forms of life it might contain.

Advocates of the doctrine of spontaneous generation pointed out that heat might kill some 'vital principle' essential to the production of life out of inanimate matter. Heating broth and sealing it away would in that case fail to produce life. Exposing heated broth to air that had likewise been heated was no better.

In 1864, however, the French chemist Louis Pasteur (1822–1895) produced the clincher. He boiled a meat broth until it was sterile, and did so in a flask with a long, thin neck that bent down, then up again, like a horizontal *S*. Then he neither sealed it off nor stoppered it. He left the broth exposed to cool air.

The cool air could penetrate freely into the vessel and bathe the broth. If it carried a 'vital principle' with it, that was welcome. What did not enter, however, was dust and microscopic particles generally. These settled at the bottom of the curve of the flask's neck.

As a result, the broth did not breed microorganisms and did not show any signs of life. Once Pasteur broke off the swan-neck, however, and allowed dust and particles to reach the broth along with the air, microorganisms made their appearance at once.

With that, the notion of 'spontaneous generation' seemed dead, once and for all.

Origin of life?

Once it was clearly established that spontaneous generation did *not* take place and that all life (as far as human beings were able to observe) came from previous life, it became very difficult to decide how life originated on Earth – or on any other planet.

The changeover was rather like the one that took place in

the theories concerning the origin of planetary systems. As long as one clung to an evolutionary theory such as Laplace's nebular hypothesis, it was easy to suppose that planetary systems were common and that every star was accompanied by one. The nebular hypothesis, in a way, preached the spontaneous generation of planets.

A catastrophic theory of planetary formation, however, involved an event that was so rare that planets themselves had to be regarded as excessively rare, and it became tempting to think that our own planetary system was not to be duplicated elsewhere.

In the same way, the defeat of spontaneous generation and the new suggestion that life came *only* from previous life, which came *only* from still earlier life and so on in an endless chain, made it seem that the original forms of life couldn't possibly have arisen save through some miraculous event. In that case, even if habitable planets were as plentiful as the stars themselves, Earth might yet be the only one that bore life.

Even as Pasteur was knocking the pins out from under spontaneous generation, however, the situation was being eased a little bit. In 1859, the English biologist Charles Robert Darwin (1809–1882) published a book for which *The Origin of Species* is the commonly used title.

In it he presented exhaustive evidence in favour of an evolutionary theory in which the various species of living things were not separate and distinct from the beginning. Instead, under the pressure of increasing populations and of natural selection, all living things gradually changed; new, and presumably more suitable, species developed from old. In this way, several different species might have a common ancestor and, if one went back far enough, all life on Earth may have developed from a single very primitive ancestral form of life.

The theory met with much opposition, but biologists came to accept it in time.

What it meant was that one no longer had to account for the separate creation of each of the millions of species of living things known. Instead, it would be sufficient to account for the creation of any form of life, however simple. This original simple form, produced by spontaneous generation, could then by evolutionary processes give rise to all other forms of life, however complex – even human beings.

Of course, if spontaneous generation were really impossible, the production of one form of life was as much a miracle as the production of a million forms.

On the other hand, all that biologists had done was to show that known forms of life could not be generated spontaneously in the short periods of time available in the laboratory. Suppose we dealt with a very simple form of life, much simpler than any known, and suppose we had long periods of time and a whole planet at our disposal; might not that very simple form of life finally be generated?

The key lay in that phrase *long periods of time*. The hit-or-miss random processes of evolution took a long time (even the evolutionists admitted that) and the question was whether there was time enough for the simple life form to be generated and all the myriad of complex life forms to be developed afterwards.

In Darwin's time, scientists had abandoned the notion of a planet that was no more than 6,000 years old and spoke freely of Earth's age as being in the millions of years, but even that didn't seem long enough for evolution to do its work.

In the 1890s, however, radioactivity was discovered, and it was found that uranium turned to lead with almost stupefying slowness. Half of any sample of uranium would turn to lead only after 4,500,000,000 years. In 1905, the American chemist Bertram Borden Boltwood (1890–1927) suggested that the extent of the radioactive breakdown in rock would be an indication of the length of time since that rock had solidified.

Radioactive changes of all kinds have been used to measure the age of various parts of the Earth, of meteorites, and, recently, of Moon rocks, and the general consensus now is that the Earth, and the Solar System in general, is about 4,600,000,000 years old.

Hints of this vast age were already available in the early

decades of the twentieth century, and it began to appear that there was enough time for evolution to do its work, if life could somehow start spontaneously.

But could that spontaneous start take place?

Unfortunately, by the time the extreme age of the Earth came to be understood, the extreme complexity of life also came to be understood, and the chance of spontaneous generation seemed to shrink further.

Twentieth-century chemists learned that protein molecules, which are molecules peculiarly characteristic of life, were made up of long chains of simpler building blocks called amino acids. They found that every protein had to have every one of thousands of different atoms (even millions in some cases) placed just so if it was to do its work properly. Later on, they discovered that an even more fundamental type of molecule, those of the nucleic acids, were even more complicated than protein molecules. What's more, different nucleic acids and different proteins, along with smaller molecules of all kinds, intermeshed in complicated chains of reactions.

Life, even the apparently simple life of bacteria, was enormously more complicated than had been imagined in the days when the matter of spontaneous generation was being squabbled over. Even the simplest forms of life imaginable would have to be built up out of proteins and nucleic acids, and how did *those* come to be formed out of dead matter? The origin of life on Earth, despite evolution, seemed more than ever a near-miraculous event.

Some scientists gave up and, in effect, washed their hands of the problem. The Swedish chemist Svante August Arrhenius (1859–1927) published the book *Worlds in the Making* in 1908, which took up the matter of the origin of life. In the book, Arrhenius upheld the universality of life and suggested that it was a common phenomenon in the Universe.

He went on to suggest that life might be, in effect, contagious. When simple living things on Earth form spores, the wind carries them through the air to burgeon in new places. Some by the blind force of the wind might be wafted upwards high into the atmosphere and actually, Arrhenius speculated, out into space. There they might drift for millions of years through vacuum, pushed by the Sun's radiation, protected by a hard, impervious pellicle and fiercely retaining the spark of life inside. Eventually, a spore would encounter some suitable planet without life and from it life would start on that planet.

In fact, Arrhenius suggested, that was how life on Earth got its start. It was vitalized by spores from space; spores that had originated on some other world that might remain forever unidentified.

Several points can be used to argue against this notion. One can calculate how many spores must leave a world in order that even one might have a reasonable chance to meet another world in the course of the lifetime of the Universe, and the amount is preposterously high.

Then, too, it is unlikely that spores can survive the trip through space. Bacterial spores are highly resistant to cold, even extreme cold; they might also be expected to survive vacuum. It is doubtful that even the hardiest spores could exist for the length of time it would take to drift from planetary system to planetary system, but we could stretch a point and suppose that at least some could. What we do know, though, is that spores are very sensitive to ultraviolet light and other hard radiation.

They are not subjected to this on Earth, where the air forms a blanket that filters out the Sun's more energetic radiation; nor was Arrhenius, in his time, aware of the extent to which energetic radiation fills the Universe. The radiation from any star anywhere in its ecosphere would be enough to kill wandering spores that were originally adapted to life within a protective atmospheric blanket. Cosmic-ray particles would kill them even in the depths of space.

Arrhenius thought that radiation pressure would propel spores away from a star and through space. We now know the Solar wind is much more likely to do so. In either case, whatever propels the spore away from a star and towards others in the first place would repel the spore as it approached another star and prevent it from landing on a planet within the ecosphere.

All in all, the notion of Earth's having been seeded by spores from other worlds is exceedingly dubious.

Besides, of what use is it to explain the origin of life on Earth by calling upon life on other planets for help? One would have then to explain the origin of life on the other planet. And if it could form on any planet by some natural and nonmiraculous means, then it could form on Earth in the same fashion.

But how? Even as late as the 1920s, biologists were at a loss for a natural mechanism.

The primordial Earth

One objection to the spontaneous generation of life on Earth is this: If life could be formed out of nonlife in the far past, it should happen periodically in later times, even right now. Since no such formation is ever observed in the present day, ought we not to conclude that it did not happen in the far past, either?

The fallacy in this argument is plain. It surely must be that the primordial Earth in the days before life existed upon it had characteristics different from those of today. It follows, if this is so, that we cannot argue from events now to events then. What is not likely now and does not, therefore, take place, might have been quite likely then, and *did* take place.

One obvious difference between modern Earth and primordial Earth, for instance, is that modern Earth has life and primordial Earth had not. Any chemical substance that arose spontaneously on Earth today and that was approaching the level of complexity where it might be considered as protolife would surely be food for some animal and would be gobbled up. In the primordial and lifeless Earth, such a substance would tend to survive (at least, it would not be eaten) and would have a chance to grow still more complex and to become alive.

Then, too, the primordial Earth might have had an atmosphere that was different from the present one.

This was first suggested in the 1920s by the English biologist John Burdon Sanderson Haldane (1892–1964). It occurred to him that coal was of plant origin, and that plant life obtained its carbon from the carbon dioxide of the air. Therefore, before life came into being, all the carbon in coal must have been in air in the form of carbon dioxide. Furthermore, the oxygen in air is produced by the same plant-mediated reactions that absorb the carbon dioxide and place the carbon atoms within the compounds of plant tissue.

It follows, then, that the primordial atmosphere of the Earth was not nitrogen and oxygen, but nitrogen and carbon dioxide. (This sounds even more logical now than it did when Haldane suggested it, since we now know that the atmospheres of Venus and Mars are made up largely of carbon dioxide.)

Furthermore, Haldane reasoned, if there were no oxygen in the air, there would be no ozone (a highly energetic form of oxygen) in the upper atmosphere. It is this ozone that chiefly blocks the ultraviolet light of the Sun. In the primordial Earth, therefore, energetic ultraviolet radiation from the Sun would be available in much larger quantities than it is now.

Under primordial conditions, then, the energy of ultraviolet light would serve to combine molecules of nitrogen, carbon dioxide, and water into more and more complex compounds that would, finally, develop the attributes of life. Ordinary evolution would then take over, and here we all are.

What could be done on the primordial Earth, with lots of ultraviolet, lots of carbon dioxide, no oxygen to break down the complicated compounds, and no living things to eat them, could not be done on present-day Earth with its dearth of ultraviolet light and carbon dioxide and its overabundance of oxygen and life. We cannot, therefore, use today's absence of spontaneous generation as a reason to deny its presence on the primordial Earth.

This notion was supported by a Soviet biologist, Aleksandr Ivanovich Oparin (born 1894). His book, *The Origin of Life*, also published in the 1920s but not translated into English till 1937, was the first to be devoted entirely to the subject. Where he differed from Haldane was in supposing that the primordial atmosphere was heavily hydrogenated, containing hydrogen as itself, and some in combination with carbon (methane), nitrogen (ammonia), and oxygen (water).

Oparin's atmosphere makes sense in the light of what we now know about the composition of the Universe in general, and of the Sun and the outer planets in particular. Indeed, scientists now speculate that life began in Oparin's atmosphere of ammonia, methane, and water vapour (Atmosphere I). The action of the ultraviolet radiation of the Sun split water molecules, liberating oxygen, which would react with ammonia and methane to produce Haldane's atmosphere of nitrogen, carbon dioxide, and water vapour (Atmosphere II). Then, finally, the photosynthetic action of green plants produced the present-day atmosphere of nitrogen, oxygen, and water vapour (Atmosphere III).

To be sure, the talk of spontaneous generation of life on a primordial Earth, during the 1920s and 1930s, was purely speculation. There was no real evidence whatever.

Moreover, while Haldane and Oparin (both atheists) could cheerfully divorce life and God, others were offended by this and strove to show that there was no way in which the origin of life could be removed from the miraculous and made the result of the chance collisions of atoms.

A French biophysicist, Pierre Lecomte du Noüy, dealt with this very matter in his book, *Human Destiny*, which was published in 1947. By then the full complexity of the protein molecule was established, and Lecomte du Noüy attempted to show that if the various atoms of carbon, hydrogen, oxygen, nitrogen, and sulphur arranged themselves in purely random order, the chance of their arriving in this way at even a single protein molecule of the type associated with life was so exceedingly small that the entire lifetime of the Universe would be insufficient to offer it more than an insignificant chance of happening. Chance, he maintained, could not account for life.

As an example of the sort of argument he presented, consider a protein chain made up of 100 amino acids, each one of which could be any of twenty different varieties. The number of *different* protein chains that could be formed would be 10^{130}; that is, a one followed by 130 zeroes.

If you imagine that it took only a millionth of a second to form one of those chains, and that a different chain was being formed *at random* by each of a trillion scientists every millionth of a second ever since the Universe began, the chance that you would form some one particular chain associated with life would be only one in 10^{95}, which is such an infinitesimal chance it isn't worth considering.

On the primordial Earth, what's more, you wouldn't be starting with amino acids, but with simpler compounds like methane and ammonia, and you would have to form a much more complicated compound than a chain made out of 100 amino acids to get life started. The chances of accomplishing something on a single planet in a mere few billion years is just about zero, therefore.

Lecomte du Noüy's argument seemed exceedingly strong, and many people eagerly let themselves be persuaded by it and still do even today.

Yet it is wrong.

The fallacy in Lecomte du Noüy's argument rests in the assumption that pure chance was alone the guiding factor and that atoms can fit together in any fashion at all. Actually, atoms are guided in their combinations by well-known laws of physics and chemistry, so that the formation of complex compounds from simple ones are constrained by severely restrictive rules that sharply limit the number of different ways in which they combine. What's more, as we approach complex molecules such as those of proteins and nucleic acids, there is no one particular molecule that is associated with life, but innumerable different molecules, all of which are in association.

In other words, we don't depend on chance alone, but on

chance guided by the laws of nature, and that should be quite enough.

Could the matter be checked in the laboratory? The American chemist Harold Clayton Urey encouraged a young student, Stanley Lloyd Miller (born 1930), to run the necessary experiment in 1952.

Miller tried to duplicate primordial conditions on Earth, assuming Oparin's Atmosphere I. He began with a closed and sterile mixture of water, ammonia, methane, and hydrogen, which represented a small and simple version of Earth's primordial atmosphere and ocean. He then used an electric discharge as an energy source, and that represented a tiny version of the Sun.

He circulated the mixture past the discharge for a week and then analysed it. The originally colourless mixture had turned pink on the first day, and by the end of the week one sixth of the methane with which Miller had started had been converted into more complex molecules. Among those molecules were glycine and alanine, the two simplest of the amino acids that occur in proteins.

In the years after that key experiment, other similar experiments were conducted, with variations in starting materials and in energy sources. Invariably, more complicated molecules, sometimes identical with those in living tissue, sometimes merely related to them, were formed. An amazing variety of key molecules of living tissue were formed 'spontaneously' in this manner, although calculations of the simplistic Lecomte du Noüy type would have given their formation virtually no chance.

If this could be done in small volumes over very short periods of time, what could have ben done in an entire ocean over a period of many millions of years?

It was also impressive that all the changes produced in the laboratory by the chance collisions of molecules and the chance absorptions of energy (guided always by the known laws of nature) seemed to move always in the direction of life as we know it now. There seemed no important changes that pointed definitely in some different chemical direction.

That made it seem as though life were an inevitable product of high-probability varieties of chemical reactions, and that the formation of life on the primordial Earth could not have been avoided.

Meteorites

We can't, of course, be sure that the experiments set up by scientists truly represent primordial conditions. It would be much more impressive if we could somehow study primordial matter itself and find compounds that had been formed by nonlife processes and that were on the way, so to speak, to life.

The only primordial matter we can study here on Earth are the meteorites that occasionally strike the Earth. Studies of radioactive transformations within them show them to be over four billion years old and to be dating, therefore, from the infancy of the Solar System.

About 1,700 meteorites have been studied; thirty-five of them weighing over a ton apiece. Almost all of them, however, are either nickel-iron or stone in chemical composition and contain none of the elements primarily associated with life. They therefore give us no useful information concerning the problem of the origin of life.

There remains, however, a rare type of meteorite, black and easily crumbled – the 'carbonaceous chondrite'. These actually contain a small percentage of water, carbon compounds, and so on. The trouble is, though, that they are much more fragile than the other types of meteorites, and though they may be common indeed in outer space, few survive the rough journey through the atmosphere and the collision with the solid Earth. Fewer than two dozen such meteorites are known.

Carbonaceous chondrites, to be useful to us, should be studied soon after they have fallen. Any prolonged stay on the ground is sure to result in contamination by Earthly life or its products.

Two such meteorites, fortunately, were seen to fall and were examined almost at once. One fell near Murray, Kentucky, in 1950, and another exploded over Murchison, Australia, in September 1969.

By 1971, small quantities of eighteen different amino acids were separated out of the Murchison fragments. Six of them were varieties that occur frequently in the protein of living tissue; the other twelve were related to these chemically, but occurred infrequently or not at all in living tissue. Similar results were obtained for the Murray meteorite. Agreements between the two meteorites that fell on opposite sides of the world, nineteen years apart, were impressive.

Towards the end of 1973, fatty acids were also detected. These differ from amino acids in having longer chains of carbon and hydrogen atoms and in lacking nitrogen atoms. They are the building blocks of the fat found in living tissue. Some seventeen different fatty acids were identified.

How did such organic molecules happen to be found in meteorites? Are the meteorites the products of an exploded planet?* Are the carbonaceous chondrites part of a planetary crust that bore life once and that still carry traces of that life now?

Apparently, this is not likely. There are ways of telling whether the compounds discovered in meteorites are likely to have originated in living things.

Amino acids (all except the simplest, glycine) come in two varieties, one of which is the mirror image of the other. These are labelled L and D. The two varieties are identical in ordinary chemical properties, so that when chemists prepare the amino acids from their constituent atoms, equal quantities of L and D are always formed.

When amino acids are used to build up protein, however, the results are stable only if one group is used, either the L only or the D only. On Earth, life has developed with the use of L only (probably through nothing more meaningful than

* This is one early and dramatic theory that is not generally accepted now.

chance), so that D-amino acids occur in nature very rarely indeed.

If the amino acids in the meteorites were all L or all D, we would strongly suspect that life processes similar to our own were involved in their production. In actual fact, however, L and D forms are found in equal quantities in the carbonaceous chondrites, and this means that they originated by processes that did not involve life as we know it.

Similarly, the fatty acids formed in living tissues are built up by the addition to each other of varying numbers of 2-carbon-atom compounds. As a result, almost all fatty acids in living tissue have an even number of carbon atoms. Fatty acids with odd numbers are not characteristic of our sort of life, but in chemical reactions that don't involve life they are as likely to be produced as the even variety. In the Murchison meteorite, there are roughly equal quantities of odd-number and even-number fatty acids.

The compounds in the carbonaceous chondrites are not life; they have formed in the *direction* of our kind of life – and human experimenters have had nothing to do with their formation. On the whole, then, meteoritic studies tend to support laboratory experiments and make it appear all the more likely that life is a natural, a normal, and even an inevitable phenomenon. Atoms apparently tend to come together to form compounds in the direction of our kind of life whenever they have the least chance to do so.

Dust clouds

Outside the Solar System we can see the stars, but we have eliminated them as breeding grounds of life. Perhaps we could find breeding grounds if we could inspect the cool surfaces of the planets revolving about them.

We can't do that, but there *is* cool matter in outer space that we can indeed detect; matter in the form of thin gas and dust that fills interstellar space.

The interstellar material was first detected about the turn of the century because certain wavelengths of light from distant stars were absorbed by the occasional atoms that drift about in the vastness of space. By the 1930s, it was recognized that the interstellar medium contained a wide variety of atoms, probably some of every type of atom cooked in the interiors of stars and broadcast into space during supernova explosions.

The density of the interstellar matter is so low that it seemed natural to supose that it consisted almost entirely of single atoms and nothing else. After all, in order for atoms to combine to form a molecule, they must first collide, and the various atoms are so widely spread apart in interstellar space that random motions will bring about collisions only after excessively long periods.

And yet, in 1937, stars shining through dark clouds of gas and dust were found to have particular wavelengths missing that pointed to absorption by a carbon-hydrogen combination (CH) or a carbon-nitrogen combination (CN). For the first time, interstellar molecules were found to exist.

To be sure, CH and CN are the kind of combination that can be formed and maintained only in very low-density material. Such atom combinations are very active and would combine with other atoms at once, if other atoms were easily available. It is because such other atoms are available in quantity on Earth that CH and CN do not exist naturally as such, on the planet.

No other combinations were noted in the interstellar dust-clouds through dark lines in the visible spectrum.

After World War II, however, radio astronomy became increasingly important. Interstellar atoms can emit or absorb radio waves of characteristic lengths – something that requires far less energy than emission or absorption of visible light, and therefore takes place more readily. The emission or absorption of radio waves can be detected easily, given the radio telescopes required for the purpose, and the compounds responsible can be identified.

In 1951, for instance, the characteristic radio-wave emission by hydrogen atoms was detected, and the presence of interstellar hydrogen was thus observed directly for the first time and not merely deduced.

It was understood that next to hydrogen, helium and oxygen were the most common atoms in the Universe. Helium atoms don't cling to any other atoms, but oxygen atoms do. Should there not be oxygen-hydrogen combinations (OH) in space? This should emit radio waves in four particular wavelengths, and two of these were detected for the first time in 1963.

Even as late as the beginning of 1968, only three different atom combinations had been detected in outer space: CH, CN, and OH. Each of these were 2-atom combinations that seemed to have arisen from the chance occasional collisions of individual atoms.

No one expected that the far less probable combination of three atoms would build up to detectable level, but in 1968 the characteristic radio-wave emissions of water and ammonia were detected in interstellar clouds. Water has a 3-atom molecule, two of hydrogen and one of oxygen (H_2O) and ammonia has a 4-atom molecule, one of nitrogen and three of hydrogen (NH_3).

This was utterly astonishing, and 1968 witnessed the birth of what we now call astrochemistry.

In fact, once compounds of more than two atoms were detected, the list grew rapidly longer. In 1969, a 4-atom combination involving the carbon atom was discovered. This was formaldehyde (HCHO). In 1970, the first 5-atom combination was discovered, cyanoacetylene (HCCCN). That same year came the first 6-atom combination, methyl alcohol (CH_3OH). In 1971, the first 7-atom combination was discovered, methylacetylene (CH_3CCH).

So it went. Over two dozen different kinds of molecules have now been detected in interstellar space. The exact mechanism by which these atom combinations are formed is not as yet clear, but they are there.

And even in outer space, the direction of formation would seem to be in the direction of life.* In fact, both in meteorites and in interstellar clouds it is interesting that the carbon chains are forming and that there is no sign of complex molecules that do not involve carbon. This is evidence in favour of our assumption that life (as we know it) always involves carbon compounds.

All of this evidence – in the laboratory, in meteorites, in interstellar clouds – makes it look as though the Haldane-Oparin suggestions are correct. Life *did* start spontaneously on the primordial Earth, and all indications would seem to be that it must have started readily, that the reactions in that direction were inevitable.

It follows that life would therefore start, sooner or later, on *any* habitable planet.

When life started

But how much sooner, or later, is 'sooner or later'? When did life start on the Earth?

Our knowledge of ancient life forms upon the Earth comes almost entirely from our study of fossils – remnants of shells, bones, teeth, wood, scales, even fecal matter – that have withstood at least some of the ravages of time and have done so sufficiently to tell us something about the structure, appearance, even behaviour of the organisms of which they were once part.

Fossils can be dated in various ways, and the oldest ones

* The English astronomer Fred Hoyle (born 1915) is sufficiently impressed by this to suggest that in comets (which in some ways have the composition of interstellar clouds) compounds form that are complex enough to possess the properties of life; that the equivalent of viruses are formed; and that comets may therefore be the cause of the occasional pandemics that afflict the Earth by sending new viruses into the atmosphere. It is an interesting suggestion, but it is hard to see how it can be taken seriously.

that we can deal with easily are from the Cambrian period (so called because the rocks from that period were first studied in Wales, which in Roman times was called Cambria).

The oldest Cambrian fossils are 600 million years old, and it was tempting to assume that that was when life on Earth began, more or less. However, since we know Earth is 4,600,000,000 years old, that would mean it lay four billion years without life. Why so long? And if lifelessness continued for that long, why did life suddenly appear? Why is Earth not still lifeless?

Then, too, at the time the fossil record begins in the Cambrian, life is already plentiful, complex, and varied. To be sure, all the life of which we have a record from that period is marine; there is no freshwater life or land life. Then, too, it is all invertebrate. The earliest chordates (the group to which we belong) did not appear for another 100 million years.

Nevertheless, what does exist seems quite advanced. Thousands of species of trilobites are found in the Cambrian period; these are complex arthropods very much like the horseshoe crabs of today. It is impossible to suppose that they sprang out of nothing and split up into many species. Before the Cambrian time, there must stretch long ages of simpler life. In that case, why is there no record of it?

The most likely answer is that the simpler life was not particularly prone to fossilization. It lacked the kind of parts – shells, bones – that survive easily. And yet despite that, traces of earlier life *have* been found.

The American botanist Elso Sterreberg Barghoorn (born 1915), who in the 1960s was working with very ancient rocks, came across faint traces of carbon that, as he could demonstrate, were the remains of microscopic life.

The dim evidence of such microscopic life has now been traced back as far as 3,200,000,000 years, and it probably extended back a few hundred million years before that.

We might conclude, then, that recognizable life forms existed by the time the Earth was one billion years old.

This sounds reasonable intuitively. We can well imagine

that during the first half-billion years of Earth's history the planet may have been in a pretty unsettled state. The crust must have been active and volcanic; the ocean and atmosphere in the process of formation as the planet cooled off from the heat of its initial condensation and its components separated. The second half-billion years may well have been devoted to a slow chemical evolution – the formation of more and more complicated compounds under the lash of the Sun's ultra-violet light. Finally, a billion years after the Earth's formation, very simple little bits of life exist here and there.

The Sun's stay upon the main sequence will be some twelve billion years, and we might consider this average for Sunlike stars. That means that the Earth (and, on the average, habitable planets generally) will last twelve billion years as the abode of life. If, then, life appears on the Earth after one billion years, it does so after only eight per cent of its lifetime has elapsed.

We can assume that (by the principle of mediocrity) habitable planets in general gain life after some eight per cent of their lifetimes as habitable planets has passed.

Suppose we assume that stars have been forming at a steady rate here in the outskirts of the Galaxy, once the first flurry of star formation in the infancy of the Galaxy had passed.

This is not entirely an assumption. There is evidence that stars have been born in recent times, at least. The giant stars of spectral classes O and B must have been formed a billion years ago or less, or they would not still be on the main sequence. And if stars could form in the last billion years, they must have been forming all along and still be forming now. They must at least be doing so in the galactic regions where clouds of dust and gas (the raw material of stars) are plentiful, and those regions are precisely in the outskirts of galaxies, which, we have already decided, are the only places life can exist.

Moreover, we need not depend entirely on reason to tell us that stars are still being formed today. It is possible we are actually witnessing the process. In the 1940s, the Dutch-American astronomer Bart Jan Bok (born 1906) drew attention to certain dust clouds that were opaque, compact, isolated, and more or less spherical in shape. He suggested that these clouds (now called Bok globules) are in the process of condensing into stars and planetary systems. The evidence since then tends to show he is right. Sagan estimates that in our Galaxy, ten stars are born each year on the average.

Assuming, then, a steady rate of star formation, we can say that x per cent of the habitable planets have not yet expended x per cent of their lifetime. In other words, 50 per cent of the habitable planets have not yet expended 50 per cent of their lifetime; 10 per cent of the habitable planets have not yet expended 10 per cent of their lifetime; and so on.

This means that 8 per cent of the habitable planets have not yet expended the 8 per cent of their lifetime that they should need to form life; that is, they are less than a billion years old.

The converse is that 92 per cent of the habitable planets *are* old enough to have had life develop upon them.

That gives us our ninth figure:

9 The number of life-bearing planets in our Galaxy = 600,000,000.

Multicellular life

Though life may have come to exist on Earth early in its history, its advance was very slow for a long time.

For the first two billion years during which life existed on Earth, the dominant forms may have been bacteria and blue-green algae. These were small cells, considerably smaller than the cells that make up our bodies and those of the plants and animals familiar to us. Furthermore, the bacterial cells and the blue-green algae did not have distinct nuclei within which the deoxyribonucleic acid (DNA) molecules that controlled the chemistry and reproduction of the cells were confined.

The difference between these two kinds of cells was that the blue-green algae were capable of photosynthesis (the use of energy of sunlight to convert carbon dioxide and water into tissue components) and the bacteria were not. Bacteria, without photosynthetic ability, were forced to break down already existing organic compounds for energy (or, in some cases, to take advantage of other types of chemical changes for the purpose).

Although the blue-green algae made use of the energy of sunlight to form their tissue components, they thereafter made use of chemical changes similar to those the bacteria used. These chemical changes did not supply much in the way of energy, so that the growth and multiplication of living things – to say nothing of its evolution into various more advanced species – was extremely slow. The reason for this is that the chemical changes that yield considerable energy to living things on our Earth of today all involve the utilization of molecular oxygen, and in the early days of life on Earth there was virtually no oxygen in the atmosphere.

The blue-green algae did produce small quantities of oxygen in the course of their photosynthesis, but the sparse distribution and feeble activity of the tiny cells made these quantities very small indeed.

But even though evolution progresses very slowly, it progresses. About 1,500,000,000 years ago, when Earth had been the abode of life for over two billion years, the first cells with nuclei appeared. These were large cells of the type that exist today, with more efficient chemistries, that were capable of conducting photosynthesis at greater rates than before.

This meant that oxygen began entering the atmosphere in perceptible quantities and carbon dioxide began declining. By 700 million years ago, after Earth had been the abode of life for nearly three billion years, the atmosphere was some five per cent oxygen.

By this time, those forms of burgeoning animal life, still made up of single cells, which like bacteria made use of chemical changes rather than sunlight as a source of energy, began to develop means of making use of the free oxygen of the atmosphere. Combining organic compounds with oxygen releases twenty times as much energy for a given mass of such compounds as does the breakdown of organic compounds without the use of oxygen.

With a flood of energy at their disposal, animal life (and plant life, too) was able to move more quickly, live more briskly and efficiently, reproduce more copiously, evolve in more different directions. It could even make use of energy in what would have been a wasteful fashion by earlier standards. It evolved into organisms in which cells clung together and specialized. Multicellular organisms were developed, and rigid tissues had to be formed to support them and serve as anchors for muscles.

Such hard tissue was easily fossilized and thus by 600 million years ago, it seems (from the fossil record) that out of nowhere multicellular life, advanced and complex, was flourishing.

It was not until Earth was four billion years old, with a third of its life span gone, that such complex life forms existed.

If this, by the principle of mediocrity, is characteristic of completely Earthlike planets in general, then one-third of them are too young to have anything more than one-celled life. Conversely, two-thirds of them possess complex and varied multicellular life.

That gives us our tenth figure:

10 The number of planets in our Galaxy bearing multicellular life = 433,000,000.

Land life

However complicated and specialized a life form becomes, it doesn't interest us in connection with the subject matter of this book unless it is intelligent.

It cannot become intelligent unless it develops a large brain (or the equivalent – except that, on Earth at least, we know of no equivalent) and this, it would appear, cannot be done without the development of manipulative organs of some sort and of elaborate sense organs of considerable variety.

It is the flood of impressions entering the brain from the outside Universe, and the questing manipulative organs that respond to these impressions, which stretch the brain's resources to its capacity and beyond, and lend survival value to any increase in the brain's size and complexity. If a small brain is already sufficient to handle the coordinating needs of the information an organism collects, a larger brain is of no advantage; a larger brain would merely require the production of useless and energy-wasting highly complex tissue. If, on the other hand, the brain is being used to capacity, a larger brain can do more and is worth much more.

Viewed from this angle, the sea is ideal as an incubator of life, but is very poor as an incubator of intelligence. The most valuable and information-rich sense that we can imagine life possessing (without veering into fantasy) is that of vision. Under water, vision is limited, for water absorbs light to a far greater extent than air does. In air, vision is a long-distance sense; in water, only a short-distance sense. (To be sure, hearing is even more efficient in water than in air and can perform wonders, but the smallest sound waves used by life forms are still far longer than the tiny light waves, and therefore incapable of transmitting as much information.)

When it comes to manipulative organs, as I mentioned earlier in the book, the necessity for streamlining to allow rapid travel through the viscous medium of water eliminates almost any chance for developing a manipulative organ. What manipulation a sea organism can perform usually involves the mouth, the tail, or the full weight of the body, and it is rarely delicate in its nature.

One exception to this is the octopus and its relatives. The octopus has developed a set of sensitive and limber tentacles with which there can be fine manipulation of the environment, yet when it wishes to travel quickly it can trail them behind and be streamlined. Then, too, the octopus has an excellent eye, the closest approach to the vertebrate eye in any nonvertebrate creature.

But though we may admire the intelligence of the octopus,

it is certainly far from intelligent enough to build what we would consider to be a civilization.

There are, of course, sea animals far more intelligent than the octopus, but these – sea otters, seals, penguins – are all land creatures who had secondarily adapted to the water again. Even the whales and dolphins have land animals among their ancestry, and it is undoubtedly in the course of the period during which their ancestors inhabited the land that the cetacean brain developed.

For real intelligence of the level in which this book is interested, then, we must consider land organisms – land organisms who can make use of sight as a long-distance sense of incredible detail and richness; who can develop manipulative organs; and who live surrounded by free oxygen so that they can tame fire and develop a technology.

And yet when all life existed in the sea only, the land was an environment extremely hostile to life; as hostile as space is to us. We, at least, in conquering space can make use of our technology and devise artificial protective devices. Sea life, hundreds of millions of years ago, had to develop protection as part of their bodies through the slow course of evolution.

Consider the difficulties they had to overcome:

In the sea, organisms need not fear thirst and drought; they are always surrounded by water, the essential chemical background to life. On the land, on the other hand, life is a continual battle to avoid water loss; water must either be conserved, or it must be replaced by drinking.

In the sea, oxygen is easily absorbed from the water in which it is dissolved. On the land, oxygen must first be dissolved in the fluid lining the lungs and then absorbed, and the lungs must not be allowed to dry out in the process.

In the sea, eggs can be laid in the water and allowed to develop and hatch without care (or with minimal care) in a congenial environment. On the land, eggs must be developed that have a shell that will prevent water loss while allowing gases to pass through freely so that oxygen can reach the developing embryo.

In the sea, temperature scarcely varies. On the land, there are extremes of hot and cold.

In the sea, gravity is almost nil. On the land, it is a powerful force, and organisms must develop sturdy legs that can lift them free of the land, or else they are condemned to crawl.

It is no wonder that even after life in the sea grew energetic and complicated it took hundreds of millions of years to conquer the land.

But the conquest took place. The pressures of competition forced organisms of various sorts to spend more and more time upon the land, until such time as they could live on land more or less permanently.

About 370 million years ago, the first plants invaded the land. The land that had been lying sterile and dead for $4\frac{1}{4}$ billion years began to turn a faint green about its edges.

Animals followed the plants over the next few tens of millions of years. Insects and spiders appeared as the first true land animals about 325 million years ago. Snails and worms appeared on land. The first vertebrates to be entirely land animals, primitive reptiles, appeared 275 million years ago.

A rich land life appeared when the Earth was about 4.3 billion years old and had passed through 36 per cent of its lifetime. By the principle of mediocrity, then, we can say that 64 per cent of the habitable planets have a rich land life.

That gives us our eleventh figure:

11 The number of planets in our Galaxy bearing a rich land life = 416,000,000.

Intelligence

Even a land species is not necessarily intelligent. To this day, cattle and other grazing animals are not particularly bright.

Nevertheless, one can see a steady progression of intelligence and a steady elaboration of the brain. Mammals, which first appeared about 180 million years ago, were on the whole an advance in intelligence over the reptiles.

The order of primates, the earliest records of which date back seventy-five million years, moved towards specialization in eyes and brains. About thirty-five million years ago, the primates split into the less brainy and smaller monkeys and lemurs on one side, and the more brainy and larger apes on the other.

Some eight million years ago, a particularly brainy species developed that was the first hominid. About 600,000 years ago, *Homo sapiens* had developed, and about 5,000 years ago, human beings invented writing, so that written history began and civilization was in full bloom, in some parts of the world at any rate.

By the time civilization appeared, the Earth was 4,600,000,000 years old and had completed roughly 40 per cent of its lifetime. That means, if we follow the principle of mediocrity, that 40 per cent of the habitable planets in existence are not old enough to have developed a civilization and 60 per cent *are* old enough.

That gives us our twelfth figure:

12 The number of planets in our Galaxy on which a technological civilization has developed = 390,000,000.

In other words, one star out of 770 in the Galaxy today has shone down on the development of a technological civilization.

We can go a little bit further. Our own civilization, if we count from the invention of writing to the first venture into space, has lasted 5,000 years. If we want to be glowingly optimistic about it, we can suppose that our civilization will continue to last on Earth as long as the Earth can support life – for another 7.4 billion years – and that our level of technology will advance in all that time.*

Suppose we say, then, that the average duration of a civilization is 7.4 billion years (we'll have more to say about that later

---

* Of course our physical shapes will surely change as time passes, thanks to evolution, or to the deliberate genetic engineering introduced by human beings, but that does not affect the line of argument.

on in the book) and that space flight is reached in the first 5,000 years. That means that only $\frac{1}{1,500,000}$ of a civilization passes before space flight is developed, and all the rest of it progresses to technological levels above and beyond that. Or, to put it another way, only $\frac{1}{1,500,000}$ of the civilizations in our Galaxy are so unadvanced that they are barely at the brink of spaceflight or have not yet reached it. All the rest are beyond us.

That means that of the 390 million civilizations in our Galaxy, only 260 are as primitive as we are – an inconsiderable number. All the rest (meaning just about all of them) are more advanced than we are.

In short, what we find ourselves to have been doing is to have worked out not merely the chances of extraterrestrial intelligence but the chances of superhuman extraterrestrial intelligence.

10 Civilizations elsewhere

Our giant satellite

In a way, our speculations concerning extraterrestrial intelligence have ended on a triumphant note. Doing our best to make reasonable and conservative estimates and assumptions, we end with a Universe that may be incredibly rich in intelligence. Along with our own, 390 million sets of companions in the great adventure of learning and speculating have entered into civilization right here in our own Galaxy.

If those 390 million civilizations are spread evenly about the Population I outskirts of the Galaxy, then the distance between two neighbouring civilizations would be, on the average, about forty light-years. That is not very great as cosmic distances go.

But then there is a question that, in a way, spoils everything.

Where is everybody?

If there were indeed hundreds of millions of advanced civilizations in our own Galaxy, we should think that they might well have ventured beyond their own worlds; they might have formed alliances; they might have formed a Galactic Federation of Civilizations with emissaries sent to other galactic federations beyond the intergalactic spaces. And, in particular, they should have visited us. Why haven't they?

Where is everybody?

There are a number of possible explanations for this puzzle. It may be, for instance, that the analysis presented in this book is wrong in some key point after all, and there are, therefore, no habitable worlds except our own Earth.

Almost every stage in the analysis might hide an error arising out of our incomplete knowledge. Perhaps binaries are much more common than we think and much more influential in distorting planetary formation. In that case, there might be very few single Sunlike stars and very few planetary systems like our own Solar System.

It might be that the ecosphere is very shallow, as some calculations indicate, and it might be that almost no planets manage to be located in just the thin shell of space around a star that would make habitability possible.

It might be that, for some reason we as yet do not understand, planets with the mass of Earth form only rarely, and that in planetary system after planetary system, there are planets that are too large and others that are too small and virtually nowhere are there planets that are just right.

It might be an incredible cosmic accident that liquid water has collected on our world in appropriate amounts, or that other things are just so, and that we are, therefore, the only habitable planet in the Galaxy, or even in the Universe.

We have, however, no reason to think these things just yet. Evidence that will justify such thoughts may arrive at any time – tomorrow, for all we know. Until then, we have no choice but to stay with our line of reasoning and see if we can find an explanation for the absence of positive evidence of other civilizations elsewhere.

Perhaps it is not some error that arises out of our ignorance. Perhaps there is an error that arises out of something that is perfectly obvious but that we have been ignoring. For instance, is there something so unusual about the Sun, or its planetary system, or Earth itself, that we cannot truly make use of the principle of mediocrity?

As far as the Sun and the planetary system in general are concerned, there is nothing we know of. It may be unique in a dozen different ways, but in nothing that is obvious on the face of it. Not so in the case of the Earth. Here we have something that cannot help but be unusual and that we have so far ignored; that we must now consider as a possible answer to the problem of the whereabouts of our space visitors.

The unusual factor is Earth's satellite, the Moon.

I have already said that the Earth-Moon combination is the

nearest approach in the Solar System to a double planet because of the Moon's extraordinary size in relation to the world it circles.* The Moon has $\frac{1}{81}$, or 0.0123, the mass of the Earth. The following table gives the total satellite mass for each planet of the Solar System, excluding Pluto, in terms of the mass of the planet itself.

Earth (1 satellite)	0.0123
Neptune (2 satellites)	0.0013
Saturn (10 satellites)	0.00025
Jupiter (13 satellites)	0.00024
Uranus (5 satellites)	0.00010
Mars (2 satellites)	0.00000002
Pluto (1 satellite)	
Venus (no satellites)	
Mercury (no satellites)	

Taking the mass of every satellite relative to the mass of the planet it circles, the Moon is, so to speak, 6.5 times as massive as all the other satellites in the Solar System put together, excluding Charon.

From that point of view, the Moon is a most unusual satellite, and it makes the picture of a forming Earth utterly different from the other planets as they formed.

All the sizeable planets but Earth would seem to have formed about a central condensation point with at best several inconsiderable knots of matter at the outskirts, so small in comparison to the central condensation point that they could scarcely be thought to affect the manner in which the main planet is formed.

In connection with Earth, however, there seem to have been *two* condensations – one considerably larger than the other, but not overwhelmingly so.

* This statement must be modified in view of the discovery of Pluto's satellite, Charon, in 1978. Charon has $\frac{1}{10}$ the mass of Pluto, so that Pluto-Charon are much more nearly a double planet than Earth-Moon are. Pluto and Charon are, however, quite small bodies, and the line of argument in this section may still be valid if applied to bodies large enough to be Earthlike.

Consider Venus and Earth, then, so alike in mass and constitution, yet so different in present surface conditions. Is it possible that this present difference can, at least in part, be explained by the fact that Venus formed in one condensation and Earth in two? Did the Moon's formation somehow draw off material in a crucial way that acted to change the chemical or physical state of the Earth so as to initiate a different geological evolution as compared to Venus? Did that difference, slight to begin with perhaps, diverge until Earth became a cool planet with an ocean and a comparatively thin atmosphere, while Venus became a hot planet, with no liquid water at all, and a very thick atmosphere?

It might be that the double condensation that formed the Earth-Moon double planet is an exceedingly rare occurrence; so that in assuming that one out of every two planets in the ecosphere of a Sunlike star would be an Earthlike planet, we would be wrong. It would be an Earthlike planet only if it had a Moonlike satellite and that might virtually never happen. In the absence of a Moonlike satellite we would get only a Venuslike planet at best.

If that were so, we would have to conclude there were virtually no habitable planets in the Universe and that Earth was an incredibly fortuitous freak. Naturally, there would then be no extraterrestrial intelligences, or virtually none, and there would be no reason to be surprised that space is quiet and that we haven't heard from them.

Yet having argued in this fashion, can we find the argument compelling? Just what is the influence of the Moon's formation on that of the Earth? What could it have done in forming to decrease Earth's atmospheric density, increase its water supply, prevent a runaway greenhouse effect?

There is no reasonable answer to that as yet.

Finally, we can point out a way of rationalizing the differences between Venus and Earth that seems more probable than anything to do with the Moon.

Venus is closer to the Sun than Earth is and by a considerable amount. The process of photolysis, whereby the Sun's ultraviolet radiation breaks up the water molecules to hydrogen and oxygen would be accelerated; the hydrogen would escape rapidly thanks to the higher temperatures caused by the nearness of the Sun; the oxygen would combine with any methane present to form water and carbon dioxide. The process would continue, leading eventually to a thick atmosphere consisting chiefly of carbon dioxide, which would accelerate the greenhouse effect and bring about the Venus we know.

Many details remain to be worked out, but it is much easier to believe the difference between Venus and Earth rests in the difference in distance from the Sun than the difference in the nature and existence of a satellite.

Pending further evidence, then, there seems no way of denying the existence of many habitable, life-bearing planets. Even so, granting that, we have not yet done with the peculiarity of the Moon's existence.

Our captured satellite?

So odd is the Moon's existence as a satellite of the Earth, that some astronomers have suggested that it did not form as a satellite, but was captured by the Earth. If so, this, too, might have a conceivably fatal effect on our hope for the existence of civilizations elsewhere.

In favour of the possibility of the Moon's being a captured body, there is the fact that the Moon is as large as it is and as far from the Earth as it is. Moreover, its orbit is in a plane close to that in which the planets generally revolve about the Sun, and is considerably less close to the Earth's equatorial plane, where experience tells us a satellite is more likely to revolve. All that might lead one to believe it had been a small planet to begin with, rather than a satellite.

Then, too, the Moon is somewhat different in composition from the Earth. It has only three-fifths the density of Earth and lacks a metal core. In this, it much more closely resembles the structure of Mars. Could it be that the Moon was formed out of that portion of the original cloud of dust and gas from which Mars was formed, rather than Earth?

Further, the Moon is much more lacking than Earth in those solid elements that melt at a not too high temperature and that may, therefore, have boiled away from the Moon. Again, bits of glassy materials, formed by rocky substances that have melted and resolidified, are common on the Moon, though rare on the Earth. Both these characteristics of the Moon seem to indicate that it was at one time exposed, for a considerable period, perhaps, to temperatures higher than those to which Earth (or the Moon itself) are now exposed.

Could it be, then, that the Moon, formed in the same process that formed Mars, had for some reason a highly eccentric orbit? Perhaps it swooped in nearly as close to the Sun as Mercury does at one end of its orbit and receded almost as far as Mars does at the other end. That would account for its Mercurian surface and Martian interior.

Then, at one time, something happened that made it possible for Earth to capture the Moon at one of the latter's close approaches.

To be sure, none of these arguments for the Moon's status as a captured body is compelling. Its large size is not a convincing argument, for those satellites in the Solar System that astronomers are certain are captured are all tiny. The Moon's distance from the Earth could be the result of tidal action; the eccentricity of its orbit is not as great as that of other surely captured satellites; the inclination of its plane of revolution to its planet's equatorial plane is not as great as that of Neptune's satellite, Triton.

As for the difference in composition, it might be that the metals condensed first and that when the Moon began to condense at a distance from the primary condensation site, the cloud out of which it formed was predominantly rocky. To account for the great heat to which its surface was exposed, we need only remember that the Moon, unlike the Earth, lacks an atmosphere and an ocean to serve as a buffer against solar radiation.

Worst of all, the mechanics by which the Earth would be able to capture a body the size of the Moon are very tricky, and astronomers have not been able to suggest a credible way in which it could have been done in actual fact.

However, the arguments against the Moon as a captured satellite are not compelling, either. Astronomers have not yet been able to come to a decision in this respect. The Moon may not be a captured satellite, but it may be.

We can be justified, then, in assuming for the sake of argument that the Moon *is* a captured satellite and see where that leads us.

To begin with, *when* might it have been captured?

There is no way of telling, really. It could have been captured four billion years ago, not long after both bodies were formed and before any life appeared on Earth. It could have been captured four million years ago, not long before the first hominids appeared on Earth.

At least there is no way of telling if we consider only the Moon. Suppose, though, we consider the Earth. Is there any sharp revolution in the Earth's history that might conceivably be correlated with the capture of the Moon and blamed on that capture?

What about the appearance of land life on Earth? The land was colonized oddly later. Whereas life in the ocean began perhaps one billion years after the Earth was formed, life on land did not appear till 4.2 billion years after it was formed. If we equate the 12-billion-year lifetime of Earth as a habitable planet with the 70-year-old lifetime of a human being, sea life began when the Earth was 6 years old and land life when it was 25 years old. Why the difference?

Is it possible that the tides had something to do with the coming of land life?

The periodic progression of water up a shore and then down again would carry life with it. It would leave pools behind in which some forms of life could flourish. There would be water-soaked sands that could become hospitable to life. Adaptations would make it possible for life forms to withstand limited amounts of drying between one high tide and the next, creeping further up the shore until finally life was possible without any actual immersion in water at any time.

Could it be that in the nearly tideless ocean of a moonless Earth, the tidal transition between sea life and land life was absent, and that for three billion years land life did not develop?

Could the Moon have been captured a little before 600 million years ago and could the tides that suddenly resulted have sufficiently stirred up the forming sedimentary rock to wipe out the earlier traces of fossils and have helped make the appearance of life forms in the Cambrian rocks seemingly sudden?

And could a couple of hundred millions of years of tides have finally pushed life on to the land and made intelligence and technology possible?

To be sure, even with the Moon absent, the Earth is not entirely tideless. The Sun produces tides, too, and if the Moon were not in the sky, the tides produced by the Sun alone would be about one-third as high as that produced by the Sun and Moon together now.

One might argue that what the Sun could do wouldn't be enough and point out, in addition, that what the Moon could do in ages past is more than it could do now.

Because tidal effects are slowing the Earth's rotation, the Earth is losing angular momentum of rotation. Angular momentum cannot actually be lost; it can only be transferred. In this case it is transferred from the Earth's rotation to the Earth-Moon revolution. The Earth and Moon slowly recede from each other, make larger sweeps about their mutual centre of gravitation, and thus gain angular momentum.

If we look backwards in time, we can see that 400 million years ago when the transition from sea life to land life began, the day must have been shorter and the Moon must have been closer to the Earth. In actual fact, there is evidence from the growth rings on fossil corals of the period that, at that time, the day was about 21.8 hours long, and the Moon's period of revolution was twenty-one days (which meant that it was only 320,000 kilometres, or 200,000 miles, from Earth).

Remembering that the tidal effect varies inversely as the cube of the distance, we can see that the height of the lunar tides 400 million years ago was 1.66 what it is today and 1.44 what the lunar and solar tides together are now. With tides roughly one and a half times the height they are now, and moving up and down at a speed ten per cent greater than at present (thanks to the shorter day back then), the push towards land life could have been considerably more effective then than it would be today.

We *might* conclude, then, that the Earth, in accomplishing the very ticklish task of capturing the Moon (so difficult a task that astronomers can't figure out just how it might have happened) made it possible for land life to exist.

When we calculated how many myriads of habitable planets there were, we left out of account how few of them might have succeeded in capturing a large satellite that just happened to be there, and how few might therefore have developed land life and in that way have the kind of intelligence and technology that we are looking for.

And yet this argument in favour of Earth's being unique in possessing land life, and therefore intelligence and technology, is also not compelling. We don't need a captured Moon to explain the coming of land life. During the billions of years that life existed in the sea and not in the land, the Moon's tides, however high, probably could not have brought about a transfer of life to the land.

During most of Earth's existence after all, the Earth's atmosphere did not contain more than a very small percentage of free oxygen, if any at all. This meant that there was no ozone layer in its upper reaches, and ultraviolet radiation from the Sun could reach the Earth's surface in large quantities.

The energetic ultraviolet radiation is inimical to life since it tends to break up the complicated molecules on which life depends. This would not affect life in the ocean, however, which could drift just far enough under the ocean surface to receive enough of the ultraviolet energy without receiving too much.

On land, however, it is not that easy to escape the deadly radiation of the Sun, so the land remained dead, sterilized by sunlight.

Even at the beginning of the Cambrian era, 600 million years ago, the Earth's atmosphere was not quite five per cent oxygen. The oxygen content was now increasing rapidly, however, and an ozone layer was forming and growing denser. The ultraviolet was increasingly blocked by the forming ozone layer and by 400 million years ago it no longer reached the Earth's surface in deadly quantity. Now for the first time, fragile living tissue pushed up on the shore by the tides was not killed at once. Slowly, the land was colonized.

This is a much more persuasive argument for the delay in establishing land life than are those involving the capture of the Moon.

It would seem, then, that we must abandon the thought of the Moon's playing a crucial role in the development of civilizations. Whether a habitable planet has a large satellite, a small satellite, a captured satellite, several satellites, or no satellites should not affect, as nearly as we can judge from the evidence on hand, the development of land life, and intelligence will move on undisturbed.*

Then where is everybody?

Intelligence

Granted, then, that there are as many habitable planets as we have estimated and that all of them are filled with land life, can we really be sure that an intelligent species will inevitably arise on each of them?

Are we perhaps wrong to apply the principle of mediocrity

* I must stress that the 'evidence on hand' is fragmentary and uncertain. At any time – tomorrow, perhaps – new evidence may break the chain of logic in this book at any point.

to this phase of the calculations? Can it be that the development of intelligence on Earth is an unbelievably lucky chance, and that while the Galaxy and the Universe swarm with life, even with land life, intelligence and, hence, civilizations might be altogether absent – except here?

Are the requirements for an intelligent species all but impossible to meet? What are they?

In the first place, an intelligent species must be rather large, for it must develop a large brain; though it must not be too large, in the sense that its body must not outrun its brain.

Thus, the human being is more intelligent than its larger relative, the gorilla; and undoubtedly more intelligent than the still larger (and now extinct) Gigantopithecus, the largest primate that ever lived, as far as we know.

Nevertheless, the human being is one of the four largest primates now existing, and those four are all more intelligent than the smaller primates from the gibbon downwards. What's more, *Homo sapiens*, the species that is the brightest of the hominids, is also the largest.

Of the nonprimate mammals the most intelligent are the elephant and the dolphin, and they are large animals, too. The octopus, which is the most intelligent of the invertebrates, is among the larger invertebrates; and the crow, which may be the most intelligent of the birds, is among the larger birds.

This very largeness must be one of the reasons for the delay in the establishment of intelligence on Earth (and, presumably, on any similar planet), since it must take considerable time for the blind-chance processes of evolution to develop a species large enough to house a brain large enough for the purpose.

What makes it even more difficult is that the brain is by far the most complexly organized of all the tissues, so that it is far easier, so to speak, to develop additional mass and intricacy in any of the tissues other than the brain. Therefore, there are many more large-bodied, small-brained species than large-bodied, large-brained species.

Might not the difficulty of producing a large body with a

large brain be so great as to preclude it in almost every case?

Of course, we might argue that intelligence offers such advantages that the tendency towards it would be overwhelming. After all, it is our intelligence that grants human beings security against any form of life large enough, armed enough, vicious enough to demolish us if we were not intelligent. No mighty predator can stand against us. Indeed, we must make a special effort to avoid extinguishing the proudest and most magnificent species in existence – and despite all our efforts we may fail. The power of our intelligence is too great to soften and make mild.

Let us, however, not be misled by our pride. Our intelligence is not an all-encompassing advantage. There are disadvantages, too. Since an intelligent organism must be relatively large, it must also be relatively few in numbers. It must be long lived, to take advantage of its intelligence (for if it dies before it has learned much, its intelligence goes for nothing), and it must therefore reproduce comparatively slowly.

If an intelligent species must then compete with other species, which, not being intelligent, can afford to be small, numerous, fecund, and short lived, the intelligent species labours under a serious disadvantage. There is every reason to think that evolution hands the award (survival) to the quality of fecundity more than to anything else.

The intelligent species has young that are few and that are quite helpless until the extraordinarily complex brain, which cannot reach adequate growth even during an extended stay as foetus, develops sufficiently. If something happens to the young organism before it can in turn reproduce, it represents the loss of an enormous investment of time and effort (both biological and social).

A tiny unintelligent species can produce thousands or even millions of eggs, which will quickly hatch out myriads of young that can live independently of their parents. Most will be eaten, but the investment for any one of them is negligible, and some will surely survive.

What's more, to be short lived and fecund is to evolve at breakneck speed. The insects, which are the most familiar example of short-lived, fecund organisms, have evolved into more species than make up all noninsect organisms put together, and by any standard other than that set by our own vanity are the most successful group of organisms in the world.

Nor can humanity, at its present peak of intelligence and technology, defeat the insects. We can effortlessly destroy elephants and whales, but the insects, who consume large fractions of our food supply, defy us. We can kill them by the billions, but there are always more to replace the dead. If we use poisons, those few that happen to be able to resist the poison survive and at once breed billions of others, all equally resistant. We use brains, they use fecundity, and they win.

As a matter of fact, if you leave human beings to one side, other intelligent species are even less successful. Neither the gorilla nor the chimpanzee is a very successful species. Certainly neither can match the rat when it comes to making its way through a hostile world. Nor is the elephant as successful, to all appearances, as the rabbit; nor the whale as the herring.

Might we argue, then, that intelligence is essentially an evolutionary blind alley? Might we argue that, on the whole, the disadvantages outweigh the advantages until some critical level is reached and passed that will allow the intelligent species to assert at least some spectacular forms of domination over the world?

Perhaps that critical level is so difficult to reach, through and over the disadvantages of intelligence generally, that it was obtained by the hominids on Earth only through an extraordinary fluke which is duplicated nowhere else.

All this, however, does not carry conviction.

As we survey evolution on Earth, there does seem a trend in the direction of increasing size and complexity (occasionally overdone, to be sure, to the point of diminishing returns). What's more, increasing complexity seems almost always to involve increasing intelligence in widespread groups of living things.

Even among the insects, at least three groups – ants, bees, and termites – are social insects. Instead of growing into large and complex individuals, they remain small, but form large and complex societies; and the societies seem considerably more intelligent as a whole than are the small individual organisms that make it up.

If intelligence increases in the development of many different groups of species, and even does so in two widely different ways – the elaboration of the individual and the elaboration of the society – then we have to assume that sooner of later some developing intelligence will pass the critical level.

The weight of evidence, as presently known, therefore forces us to consider that intelligence, and sufficient intelligence to produce a civilization, is more or less an inevitable development on a habitable planet, given sufficient time.

Extinction

And still we're thrown back to the same question. If we can't find any reason to deny the development of hundreds of millions of civilizations in our Galaxy, why is everything so quiet? Why has not one of them made itself known to us?

The answer may lie in the fact that so far we have only specified that so many civilizations have come into being. We have not yet asked the question as to how long a civilization may endure once it has come into being.

This is an important point. Suppose that each civilization that comes into being endures only a comparatively short time and then comes to an end. That would mean that if we could examine all the habitable planets in the Universe, we might find that on a large number of them civilization has not yet arisen, and that on an even larger number civilization has arisen, but has already become extinct. Only on a very few planets would we find a civilization that has arisen so recently that it has not yet had time to become extinct.

The briefer the duration of civilizations, on the average, the less likely we are to encounter a world on which the civilization has come and not yet gone, and the fewer civilizations *in being* there will be now – or at any given moment in the history of the Universe.

Might it be, then, that civilizations are self-limiting, and that the reason civilizations elsewhere have not made themselves known to us is that they don't endure long enough to be heard from?

Is there reason to suppose that civilizations might be short lived? Unfortunately, judging from the one civilization we know – our own – the task of finding reason is all too easy.

Our own civilization has a dubious future, and if we can express the reason in brief it is that we find it difficult (perhaps impossible) to cooperate in solving our problems. We are too contentious a species and apparently find our local quarrels to be more important than our overall survival.

In a way, all living things must be contentious. Reproductive capacities are such that any species, reproducing freely, can in short order outrun its food supply, however plentiful.* Consequently, in the case of any species there will always be a race for food among themselves. The competition may not be direct and need not involve confrontation, and yet the survival of some will mean (and is dependent on) the nonsurvival of others. Even plants compete vigorously and remorselessly for sunlight.

The danger to civilization, then, is not just that human beings are contentious, but that they are far more contentious than other species. For this we can see several reasons, every

---

* It always yields ludicrously short intervals of time when we try to calculate how long it would take a virus, a bacterium, a pair of flies, a pair of mice, a pair of human beings, even a pair of elephants, to produce offspring equal in mass to the entire Earth – assuming free reproduction, unlimited food, and no deaths but by old age. In the case of human beings, if we start with one pair and multiply them at an overall rate of 3.3 per cent a year – easily within human capacity – the descendants of that one pair will be equal in mass to the entire Earth in 1,600 years.

single one having to do with intelligence – which is unfortunate, for it may mean that all species capable of building a civilization must be perforce overly contentious.

For instance, thanks to their intelligence, human beings are more apt than any other species to understand that competition exists. For human beings it is not just the striving for the immediate scrap of food, or the guarding of an immediate kill. For human beings, it is the working out of a long-range scheme for getting the better of others.

In other species, a quarrel over food will last until one individual succeeds in swallowing it, whereupon the other individual, disappointed, moves away to seek something else. There is no point in fighting and striving once the food is gone.

For the intelligent human beings, capable of forethought and therefore understanding what death by starvation means and how likely it might be at a given time, a quarrel over food is more likely to be violent and of long duration, and to end in serious injury and death. What is more, even if one individual is beaten and driven off without serious injury to himself, and the food is eaten by the victor, the fight may not be over.

The human being is intelligent enough to hold a grudge. The loser, remembering the injury to his own chances of survival, may then strive to kill the winner by trickery, or from ambush, or by rallying friends – if he cannot do it by main force. And the loser may do this not for any direct good it will do him, or for any increase in the chance of his survival, but out of sheer anger at the memory of the harm done him.

It is not likely that any species other than the human being kills for revenge (or to prevent revenge, since dead people tell no tales and plot no ambushes). This is not because human beings are more evil than other animals, but because they are more intelligent than other animals, and can remember long enough and specifically enough to give meaning to the concept of revenge.

Furthermore, to other species there is little else but food, sex, and the security of the young over which to quarrel. In the case of the human being, however, with his intelligent capacity for foreseeing and remembering, almost any object is liable to set off a spasm of competitive acquisitiveness. The loss of some ornament, or the failure to seize one, may set up a grievance that will lead to violence and death.

And, as civilization approaches and is achieved, human beings develop a more and more materialistic culture, one in which the possession of any number of different things is held to be of value. The development of hunting makes stone axes, spears, bows, and arrows valuable. The coming of agriculture gives land a much greater value than ever before. Rising technology multiplies possessions, and almost anything – from herds of animals, to pottery, to bits of metal – can be equated with economic well-being and social status. Human beings will then have reasons without number to attack, defend, maim, and kill.

Furthermore, the advance of technology cannot help but increase the power of the individual human being to commit effective violence. It is not just a matter of choosing to manufacture swords rather than ploughshares. To be sure, some products of technology are designed to kill, but almost *any* product can be used to kill if the anger or fear is there. A good heavy pot, ordinarily used for most peaceful purposes, can be used to crush a skull.

This continues without limit. Human beings now have at their disposal a series of weapons deadlier than they have ever had, and they still strive for a further intensification of deadliness.

We can conclude that it is impossible for any species to be intelligent without coming to understand the meaning of competition, to foresee the dangers of losing out in competition, to develop an indefinite number of material things and immaterial abstractions over which to compete, and to develop weapons of increasing power that will help them compete.

Consequently, when the time comes where the weapons the intelligent species develops are so powerful and destructive that they outstrip the capacity of the species to recover and rebuild – the civilization automatically comes to an end.

Homo sapiens has, it would seem, run the full gamut and now faces a situation whereby a full-scale thermonuclear war could end civilization – perhaps for ever.

Even if we avoid a thermonuclear war, the other concomitants of a developing technology that have been allowed to expand without sufficiently intelligent and thoughtful guidance could do us in. An endlessly expanding population combined with dwindling reserves of energy and material resources would inevitably bring about a period of increasing starvation, which might lead to the desperation of thermonuclear war.

The pollution of the environment may diminish the viability of the Earth – by poisoning it with radioactive wastes from our nuclear power plants, or with chemical wastes from our factories and automobiles, or with something as unremarkable as the carbon dioxide from our burning coal and oil (which may induce a runaway greenhouse effect).

Or civilization may just break down in internal violence without the thermonuclear horror, as the constraints of civilization simply fall apart under the strains of increasing populations and the decline in living standards. We see this already in the rising tide of terrorism.

Well, then, suppose that that is how it always is on any world. A civilization arrives, technological advance accelerates until it reaches the nuclear bomb level, and then civilization dies with a bang, or possibly with a whimper.

Let us further take ourselves as average and say that on every habitable planet with a total potential life-bearing duration of twelve billion years, an intelligent species comes into being after 4,600,000,000 years, and in the course of 600,000 years builds a civilization slowly and ends it quickly – ruining the planet, in the process, to the point where no further civilization can arise upon it.

Since 600,000 is $\frac{1}{20,000}$ of twelve billion, we can divide the 650 million habitable planets in our Galaxy by 20,000 and find that only 32,500 of them would be in the 600,000-year period in which a species the intellectual equivalent of *Homo sapiens* is expanding in power.

Judging by the length of time human beings have spent at different stages in their development and taking that as the average, we could suppose that 540 habitable planets bear an intelligent species that, at least in the more advanced parts of the planet, are practising agriculture and living in cities.

In 270 planets in our Galaxy, intelligent species have developed writing; in twenty planets modern science has developed; in ten the equivalent of the industrial revolution has taken place; and in two nuclear energy has been developed, and those two civilizations are, of course, near extinction.

Since our 600,000 years of humanity occur near the middle of the Sun's lifetime, and since we are taking the human experience as average, then all but $\frac{1}{20\,000}$ of the habitable planets fall outside that period, half earlier and half later. That means than on about 325 million planets no intelligent species has as yet appeared, and on 325 million planets there are signs of civilization in ruins. And nowhere is there a planet with a civilization not only alive but substantially farther advanced than we are.

If all this is so, then even though our earlier analysis of hundreds of millions of civilizations arising in our Galaxy is correct, it is no wonder that we haven't heard from them.

Cooperation

Yet this analysis, while depressing, is perhaps not completely compelling. Contentiousness is not the only factor to be considered in human beings. There is also an element of cooperation and even selflessness.

If the intelligence of a human being makes it possible for him to remember grievances and to labour to avenge them, it also makes it possible for him to sympathize with the feelings of others, to understand and forgive. Even with a completely hard heart, a human being may appreciate, for purely selfish motives, the advantages of cooperation.

After all, though an instant blow may fell a competitor and make it possible for you to eat all the food immediately present, sharing the supply and combining talents in the search for additional food may improve the long-term chances of fending off starvation.

There are countless examples in human history of unselfish devotion to family, friends, tribe, and even to abstract ideals. Countless men and women have placed any number of considerations ahead of immediate satisfaction of desires – even ahead of life.

And if the unselfish have always represented a minority in human history, their influence has been out of proportion to their numbers.

Even that most contentious of all human activities, organized warfare, could not be carried on at any level beyond the free-for-all mêlée were it not certain that soldiers would defend each other and routinely risk their lives on behalf of each other.

The result is that, on the whole, the political units of humanity (societies within which violence is placed under severe constraint and is visited with organized punishment) have tended to increase in size and population with time.

The hunting tribes of a few hundred individuals gave way to farming communities, to city-states, to empires of increasing extent. One-sixth of the land area of the world is now under the centralized rule of the Soviet government in Moscow. One-fifth of the world's population is under the rule of the Chinese government in Peking. One-third of the world's wealth is under the control of the American government in Washington.

One might suppose that the natural development is towards a political unit that will include the entire planet and all its population and wealth.

There seems precious little sign of this at the moment. The nations of the world recognize no law higher than their own will and may freely go to war with each other if they choose – and some do choose. What's more, the inner constraints may fail, and civil war or anarchic terrorism at various levels can occur.

It remains a clearly visible fact, though, that since the coming of the nuclear bomb, there has been a growing reluctance to chance war. There have been no wars between major powers since 1945; and no minor war has been allowed to embroil the major powers in active combat.

Again, there is increasing appreciation of the fact that over-population, pollution, resource depletion, and human aliena-tion are dangers that affect the entire globe, and that the solutions will have to be undertaken on a global scale. The thought seems to go against the grain, and one can almost hear the grinding of collective teeth in frustration as the peoples of the world face the annoying necessity of having to forget their grievances and suspicions in order that they might learn to co-operate.

Humanity may fail. The forces of violence may overcome those of cooperation; or else we have waited too long and even though we attempt cooperation with all our heart, we can no longer present civilization from collapsing under the gathering pressures. However, even if we lose out, it will not be an in-evitable or unopposed loss; we will put up a fight.

Either way, it may be a narrow squeak. We may collapse, having almost saved ourselves. We may survive, after suffer-ing agonies.

From this we might deduce (on the principle of mediocrity) that, on the whole, it is a narrow squeak for all civilizations. Through unpredictable accidents of history, or temperament, or even biology, some civilizations may have less chance than ours does, and some may have more.

If we view our own case as near the balance point, and think of ourselves as equally likely to fail or to survive, then we might suppose that half the civilizations that are established in the Galaxy will survive the kind of crisis we face today.

Of course, the present kind of crises is not the only deadly crisis that may face a civilization. There may be external dan-gers – a supernova may explode within a few light-years of a civilization and radiation may seriously damage the gene pool. An asteroid may collide with the planet. The star it circles may have a spasm of instability.

There may be internal dangers, too, that we can't easily predict since we have not yet reached the stage of civilization where we will be encountering them. For that matter, consider a civilization that has solved all its problems and reached a mild and secure plateau of security; that civilization may then fizzle to destruction out of sheer boredom.

It may be that sooner or later any civilization will come to an end no matter how many problems it solves.

In that case, what would the average duration of each civilization be?

For this question, we have no logical answer and no way of making any reasonable guess. We absolutely don't know and can't say.

We might argue that, from the fact that we have not been visited by any advanced civilization, the duration of civilizations *must* be short.

Before reaching that disheartening conclusion, we might make the experiment of assuming long duration and then seeing whether there can remain any logical reasons for our not having heard from our intellectual cousins among the distant stars. If there remains no reason, no reason at all, why we should not have heard from them, then we will be forced back to the short-duration-of-civilization hypothesis.

In pursuit of this experiment, let us say that the average civilization endures one million years before, for one reason or another, it comes to an end. Why a million years? Because it is a nice round figure and is both a long one in human terms and a short one in planetary terms.

Furthermore, is it fair to make the assumption, as I have been doing, that once a civilization comes to an end, it is a once-and-for-all collapse and that never again does a civilization appear on that planet?

Perhaps not. Even if humanity were to blow itself up and contaminate the land, water, and air with radioactivity, that radioactivity will dwindle with time. Some life may survive. As millions of years pass, the Earth may heal itself and geologic processes may reconcentrate its resources while evolutionary processes spread life outwards in new and flourishing varieties.

Eventually another intelligent species may arise and develop a civilization.

All the more would this be true if a humane, long-lived society ended its existence not in violence, but because of some social equivalent of old age.

We might easily suppose, then, that within a billion years, a second civilization would come into existence and live out its average lifetime of a million years. There might be, in short, second-generation civilizations, third-generation civilizations, and so on, up to perhaps tenth-generation civilizations before the planet's star leaves the main sequence.

We have no evidence that this can be so. On Earth, there seems no doubt that our present civilization is a first-generation one. There are no signs whatever of an earlier, prehuman civilization,* and from what we know of the evolutionary history of life on the planet, we can't see which prehuman living species could possibly have supported such a civilization.

Nevertheless, it is intuitively easy to believe that such a succession of generations could exist. It might even be that a dying civilization might provide for its own succession, either by the genetic engineering of some near-intelligent species, or by the creation of artificial intelligence.

Counting in all the successive civilizations existing on a planet, we might suppose that the average total duration of civilization upon a planet, during the course of its star's stay on the main sequence, is perhaps ten million years.

This is a conservative enough estimate. It means that civilization would be present on a planet like Earth for only $\frac{1}{740}$ of the time the planet will endure as a home for life after the first civilization has arisen. That means that only one star out of 570,000 shines down on civilization that is now existing.

Remembering our calculation that 390 million civilizations have come into being, we now have a thirteenth figure.

13 The number of planets in our Galaxy on which a technological civilization is now in being = 530,000.

* Even if we accept tales such as that of Atlantis, we would be dealing with only a slightly earlier version of human civilization.

Exploration

Even a consideration of the mortality of civilizations leaves us with over half a million of them now existing in our own Galaxy. We must, therefore, still ask the question: Where is everybody?

And yet, just because these half-million advanced civilizations are in our own Galaxy, let us not overestimate their closeness to us. They are not our next-door neighbours by any means.

Here on the outskirts of the Galaxy (where we have decided the civilizations must exist) the distance between two neighbouring stars that are not connected gravitationally in the form of multiple-star systems is about 7.6 light-years.

If only one star out of 570,000 shines down upon an advanced civilization now existing, the average separation of two civilizations is 7.6 light-years multiplied by the cube root of 570,000. This comes to about 630 light-years.

This is a long distance, and it may be that of all the reasons I have so far advanced as possibly explaining the lack of visits from other civilizations the impracticality of negotiating such distances is the most nearly compelling.* It may well be that every civilization, no matter how advanced, is isolated in its own planetary system and that visits among them are out of the question.

Nevertheless, it is possible to take the view that interstellar travel seems difficult to us only at our present level of technology. A hundred years ago it might have seemed to us that reaching the Moon was a matter of insuperable difficulty; that jet planes and television were mad fantasies. Yet such things are now so common we give them no thought.

Give us another hundred years – or another thousand of the prospective million-year-existence of our civilization – and might not interstellar travel become commonplace and easy?

* I will discuss the difficulties of interstellar travel in some detail in the next chapter.

We'll discuss the pros and cons later, but for now let us assume that interstellar travel is a reality for the half-million civilizations of the Galaxy and that travelling from planetary system to planetary system offers no difficulties. If that is so, why haven't they visited us?

Can it be that as one civilization after another ventures out into space, there is intersection and conflict? Even granted that each civilization, in order to get out into space, develops a planetwide political unit, might there not nevertheless be wars among the worlds?

If we want to wax dramatic, we can imagine civilizations killing each other off with devices that explode whole planets or induce stars to leave the main sequence.

Yet that seems wrong to me. Civilizations that had managed to suppress undue violence on their home worlds would have learned the value of peace. Surely they would not forget it lightly, once off their planet.

Besides, it is not likely that the struggle would be so even that, like the fabled Kilkenny cats, the various civilizations would destroy each other until none was left. Those that were more advanced might win out and establish sway over broader and broader sections of the Galaxy. Indeed, the oldest civilizations, intent on imperial growth, might take over scores, hundreds, thousands of habitable planets before those could develop native civilizations, aborting those civilizations for ever.

The half-million habitable worlds might all bear civilizations indeed, but all of those civilizations might belong to any of but a dozen different 'Galactic nations', so to speak, maintaining an uneasy peace among themselves. Perhaps the oldest or the mightiest might have managed to take over all the worlds – aborting those civilizations not yet begun, destroying or enslaving those that had a late start – and established a 'Galactic Empire'.

But if that is so, why haven't we been aborted, taken over, enslaved, destroyed? Where are these Galactic Imperial horrors?

Perhaps they are on their way. The Galaxy is so huge that they just haven't got to us yet.

Surely, that is not very likely. The Galaxy was formed fifteen billion years ago. Really large stars don't shine for very many million years before exploding, so that by the time the Galaxy was a billion years old or so, there must have been a growing number of second-generation Sunlike stars in the outskirts. Add another four billion years for civilizations to develop, and it is possible that some of them have been out in space and expanding now for ten billion years.

The Galaxy is about 315,000 light-years in circumference, so to go from any point to the antipodes, even the long way round at the very rim, in either direction, will be a little over 150,000 light-years. That means an expanding civilization would have to travel (on the average) just about the distance from the Earth to the Sun every year, no farther than that, in order to make it around the Galaxy in ten billion years.

That's just one civilization; as others are added, the rate of colonization from a growing number of nuclei grows. Even supposing no very great speeds, every corner of the habitable portions of the Galaxy must have been thoroughly explored – provided there has been the development of a practical method for interstellar voyaging.

Then why haven't they come here?

Can it be they have just overlooked us – somehow missed us in the crowds of stars?

Not very likely. Our Sun is, of course, a Sunlike star, and I doubt if in ten billion years of looking, a single such star anywhere in the Galaxy would have been overlooked.

Well, then, if interstellar travel is a practical possibility, we must have been visited; and since Earth has not been taken over and settled and our own independent civilization has in no way been interfered with, it cannot have been by Galactic Imperialists.

Civilizations expanding outward may be far more benign. They may, on principle, allow all habitable planets to develop life in their own way. They may, on principle, establish their bases and seek their resources in those planetary systems that lack habitable planets, making use instead of Marslike or Moonlike worlds.

The different civilizations may have formed a Galactic Federation and our planetary system may be a ward of the Federation, so to speak, until such time as a native civilization appears and advances to the point where it qualifies for membership.

Starships may have us under observation, for all we know. The Austrian-born astronomer Thomas Gold (born 1920) has suggested, probably in jest, that the first observation vessels may have landed on Earth when it was a new and still sterile planet, and that from the bacterial content of the garbage or wastes left behind, life on Earth began. This is a kind of reincarnation of Arrhenius's suggestion of the seeding of Earth from extraterrestrial spores.

Is all this possible? Could we imagine civilizations so concerned with other civilizations, and not 'taking over'?

Perhaps we might reason that half a million civilizations would approach the Universe in half a million different ways, produce half a million sets of cultures, half a million lines of scientific developments, half a million bodies of arts and literatures and amusements and varieties of communications and understandings. Some of all these may be capable of transmission and reception across the gap between intelligent species and, however small the portion so transmitted and received, each species is the better and wiser for it. In fact, such cross-fertilization may increase the life expectancy of each civilization that participates.

Visits

And if extraterrestrial civilizations have visited Earth and have, on principle, left us to develop freely and undisturbed, might they have visited Earth so recently that human beings had come into existence and were aware of them?

All cultures, after all, have tales of beings with supernormal powers who created and guided human beings in primitive days and who taught them various aspects of technology. Can such tales of gods have arisen from the dim memory of visits of extraterrestrials to Earth in ages not too long past? Instead of life having been seeded on the planet from outer space, could technology have been planted here? Might the extraterrestrials not merely have allowed civilization to develop here, but actually helped it?*

It is an intriguing thought, but there is no evidence in its favour that is in the least convincing.

Certainly, human beings need no visitors from outer space in order to be inspired to create legends. Elaborate legends with only the dimmest kernels of truth have been based on such people as Alexander the Great and Charlemagne, who were completely human actors on the historical drama.

For that matter, even a fictional character such as Sherlock Holmes has been invested with life and reality by millions over the world, and an endless flood of tales is still invented concerning him.

Secondly, the thought that any form of technology sprang up suddenly in human history, or that any artifact was too complex for the humans of the time, so that the intervention of a more sophisticated culture must be assumed is about as surely wrong as anything can be.

This dramatic supposition has received its most recent reincarnation in the books of Erich von Däniken. He finds all sorts of ancient works either too enormous (like the Pyramids of Egypt) or too mysterious (like markings in the sands of Peru) to be of human manufacture.

Archeologists, however, are quite convinced that even the pyramids could be built with not more than the techniques available in 2500 BC, plus human ingenuity and muscle. It is a mistake to believe that the ancients were not every bit as in-

* This was the central motif of the science fiction movie
   2001: A Space Odyssey.

telligent as we. Their technology was more primitive, but their brains were not.

Then, too, all that von Däniken finds mysterious and therefore suggestive of extraterrestrial influence archeologists are convinced they can explain, much more convincingly, in a thoroughly earthly manner.

The conclusion, therefore, is that while there is nothing inconceivable about visits to Earth by extraterrestrial civilizations in the past, even in the near past, there is no acceptable evidence that it has happened, and the evidence deduced for the purpose by various enthusiasts is, as far as we can tell, utterly worthless.

Yet even the visits of ancient astronauts are not the most dramatic suggestions of the sort. There are endless reports of Earth being visited by extraterrestrial civilizations *now*.

Such reports are usually based on the sighting of something that the sighters cannot explain and that they (or someone else on their behalf) explain as representing an interstellar spaceship – often by saying 'But what else can it be?' as though their own ignorance is a decisive factor.

As long as human beings have existed, they have experienced things they could not explain. The more sophisticated a human being is, the more widely experienced, the more likely he or she is to expect the inexplicable and to greet it as an interesting challenge to be investigated soberly, if possible, and without jumping to conclusions. The rule is to seek the simplest and most ordinary explanation consistent with the facts and to allow one's self to be driven (with greater and greater reluctance) to the more complex and unusual when nothing less will do. And if one is left with no likely explanation at all, then it must be left there; the sophisticated observer has usually learned to live with uncertainty.

Unsophisticated human beings with limited experience are impatient with puzzles and seek solutions, often pouncing on something they have vaguely heard of if it satisfies an apparently fundamental human need for drama and excitement.

Thus, mysterious lights or sounds, experienced by people

living in a society in which angels and demons are common-place beliefs, will invariably be interpreted as representing angels and demons – or spirits of the dead, or whatever.

In the nineteenth century, they were described as airships on occasion. In the days after World War II, when talk of rocketry reached the general public, they became spaceships.

Thus began the modern craze of 'flying saucers' (from an early description in 1947) or, more soberly, 'unidentified flying objects', usually abbreviated as UFOs.

That there are such things as unidentified flying objects is beyond dispute. Someone who sees aircraft lights and has never seen aircraft lights before has seen a UFO. Someone who sees the planet Venus, with its image distorted near the horizon or by a mist, and mistakes it for something much closer, has seen a UFO.

There are thousands of reports of UFOs each year. Many of them are hoaxes; many of them are honest, but capable of a prosaic explanation. A very few of them are honest and entirely mysterious. What of these?

The honestly mysterious sightings are mysterious usually only in that information is insufficient. How much information can someone gather who sees something he cannot understand and sees it without warning and briefly – and grows excited or frightened in the process?

Enthusiasts, of course, consider these mysterious sightings to be evidence of extraterrestrial spaceships. They also consider sightings that are by no means mysterious, but are clear mistakes or even hoaxes, to be evidence of extraterrestrial spaceships. Some of them even report having been on board extraterrestrial spaceships.

There is, however, no reason so far to suppose that any UFO report can represent an extraterrestrial spaceship. An extraterrestrial spaceship is not inconceivable, to be sure, and one may show up tomorrow and will then have to be accepted. But at present there is no acceptable evidence for one.

Those UFO reports that seem to be most honest and reliable report only mysterious lights. As the reports grow more dramatic, they also grow more unreliable, and all accounts of actual 'encounters of the second or third kind' would seem utterly worthless.

Any extraterrestrials reported are always described as essentially human in form, which is so unlikely a possibility that we can dismiss it out of hand. Descriptions of the ship itself and of the scientific devices of the aliens usually betray a great knowledge of science fiction movies of the more primitive kind and no knowledge whatever of real science.

In short, then, once we allow the practicality of easy interstellar travel, we are forced to speculate that Earth is being visited or has been visited, is being helped or at least left alone by a Federation of benevolent civilizations.

Well, perhaps, but none of it sounds compelling. It seems safer to assume that interstellar travel is not easy or practical.

The final conclusion I can come to at the end of the reasoning in this chapter, then, is that extraterrestrial civilizations *do* exist, probably in great numbers, but that we have *not* been visited by them, very likely because interstellar distances are too great to be penetrated.

11 Space exploration

The next targets

If the key to the paradox of the existence of many civilizations in a Universe in which to all appearances we are alone, rests with the presumed difficulty of space exploration, let us examine that problem more closely.

After all, human beings managed to place the first object in orbit, thus initiating the 'Space Age', only on 4 October 1957. Before the Space Age was a dozen years old, human beings stood on the Moon. That is a rather promising beginning. Surely we can go farther now.

In a way, we already have. Instruments have been soft-landed on the surface of Venus and Mars, and photographs and other data have been sent back to Earth. Probes have, without landing, skimmed by the surfaces of Mercury and Jupiter and have, again, returned photographs and other data. As I write, probes are on the way to Saturn and beyond.

This far penetration of human instruments without the involvement of human beings themselves does not, however, have the glorious ring of accomplishment that we associate with the mystique of exploration. Can human beings *themselves,* as distinct from their inanimate instruments, move to worlds beyond the Moon?

Unfortunately, the Moon is not a particularly hopeful precedent. It is so close to Earth that it can't help but give us a false confidence; it lures us on to underestimate the distances involved in space exploration.

The Moon, after all, is so close to Earth that it takes only three days to reach it, as compared with the seven weeks it took Columbus to cross the Atlantic Ocean.

In reaching the Moon, we have made only the most microscopic dent in the true vastness of space. Indeed, we have not really left Earth, since the Moon is as much a slave to Earth's gravitational influence as an apple on a tree – something Isaac Newton perceived three centuries ago.

To be sure, there are small bodies that occasionally come to within a few million kilometres of the Earth, ten to fifty times the distance of the Moon – an occasional asteroid or comet. The nearest sizeable body other than the Moon, however, is the planet Venus.

Even when Venus is at its closest to Earth, it is 40 million kilometres (25 million miles) away in a straight line, and is 105 times the distance of the Moon.

We cannot expect a space vessel to move straight across the gap between the planetary orbits. The most economical route for a space vessel to follow is an elliptical orbit of its own that begins at Earth and intersects the orbit of Venus just as that planet approaches the intersection point.

The probes that we have sent out to Venus take seven months to cover the distance. Those probes, however, are given one burst of acceleration at the start of their journey and are then allowed to coast the rest of the way. Time is of little importance to an inanimate object.

In the case of a manned vessel, time *is* of importance. The journey must be carried through quickly, and the easiest way of doing that is to build up greater speeds.

Human beings have more than once cancelled distance by increasing speed. I have already said that the astronauts take three days to reach the Moon, while Columbus took seven weeks to cross the Atlantic, despite the fact that the distance to the Moon is nearly eighty times the width of the Atlantic.

It's just that the astronauts travel at an average speed 1,300 times that of Columbus. Well, increase that speed by another factor of seventy and it will take only three days to reach Venus.

One way to gain the necessary speed is to place a spaceship under seventy times the acceleration of a Moon rocket, using rocket engines with seventy times the capacity for thrust. Even if we build such large engines and are willing to expend so much fuel, it remains true that the human body can only endure so much (and not very much) acceleration. The acceleration required to send the vessel on its way to Venus at a speed that would make the voyage short work would kill the astronauts at once.

The alternative is to use an acceleration no higher than that required to launch a vessel to the Moon, but then to use further acceleration at a bearable level for a prolonged period. In this way, the vessel would go faster and faster till the halfway point was reached. After that the rocket exhaust could be aimed in the other direction and a prolonged and gradual deceleration could reduce the vessel's speed for the tryst with Venus.

It would take time to accelerate and decelerate, so the voyage would take considerably more than three days. Worse yet, acceleration and deceleration requires the expenditure of energy, and we can make the general rule that to decrease the time required for any voyage means an increase in energy expenditure. (For that matter, if the astronauts move at an average speed 1,300 times that of Columbus, their total energy expenditure is far more than 1,300 times that of Columbus.)

We don't know of any way of uncoupling time lapse and energy expenditure, and if our understanding of the laws of nature is correct there is no conceivable way. Between the demands of the human body where acceleration is concerned and the demands of the human economy where energy expenditure is concerned, our first manned flights to Venus (if any) are going to take at best four months.

Already men have remained in space for almost that long, but that has been in space stations such as Skylab, in Earth's immediate neighbourhood, with rescue at short notice possible. To spend 120 days in space in cramped quarters, with every moment taking you farther from home, is a psychological hazard indeed.

Worse yet, having arrived in the neighbourhood of Venus, there would be no chance of a landing in view of the planet's almost red-hot surface temperature. Any exploration of the surface would have to be carried out by unmanned probes launched by the space vessel, which would itself remain in orbit about Venus and would then launch itself on another four-month journey back to Earth.

Since exploration of Venus's surface would have to be carried out by an unmanned probe, that probe might as well travel all the way from Earth – as several probes have indeed already done. The benefits achieved by having the probe launched from, and the signals received by, a manned mother ship would scarcely justify the traumatic experience of over eight continuous months in space.

Mercury, the planet nearest the Sun, is farther from us than Venus, being never closer to us than eighty million kilometres (50 million miles) or twice Venus's closest approach.

Mercury would at least offer a landfall to the long-distance astronauts, for one can visualize them as landing on the night side of Mercury and being able to explore the surface for several weeks before the approach of sunrise makes it absolutely necessary to leave.

The flight to Mercury, however, would carry the astronauts to a distance no farther from the Sun than 65 million kilometres (40 million miles). Solar radiation would be over four times as concentrated at that distance as it is in the neighbourhood of the Earth. For what might be gained in a manned voyage to Mercury, over an unmanned probe, the price paid in risking the effects of the greater radiation may prove too high.

Since voyages in the direction of the Sun offer no suitable target, what about voyages away from the Sun?

The nearest planet to Earth in the direction away from the Sun is, of course, Mars. It is, at its closest, some 58 million kilometres (36 million miles) away, closer than any other planet but Venus. Travelling Mars-wards means steady progress in the direction of decreasing intensities of Solar radiation. Furthermore, Mars is a cold world that can be explored for indefinite periods even with the Sun in the sky (provided there is some protection against the Solar ultraviolet other than Mars's thin and ineffective atmosphere).

Nevertheless, the round trip to Mars would certainly take more than a year of travel time. Even though that will be broken, for a shorter or longer time, by a landing on a planet that next to Earth itself is the most comfortable in the Solar System, the task would surely stretch human endurance to the limit.

And beyond Mars? To reach the larger asteroids, or the satellites of the giant planets, would mean crossing the much greater spatial gaps of the outer Solar System, and the voyages would take years and even decades one way. Manned voyages of such lengths do not seem practical at the moment.

Beyond the Moon, then, we are left with only Mars as a sizeable target and that only as a borderline possibility.

Space settlements

In a practical sense, then, our initial triumphs in space do not seem to count for much. It looks as though we will be confined to the Earth-Moon system for the foreseeable future.

That may be true, however, only because I have been assuming so far that Earth itself is the base to be used for space exploration. Is there an alternative?

If we are to be confined to the Earth-Moon system, it would seem that the Moon is the only possible alternative. Suppose we establish an elaborate base on the Moon, one where it is possible to build space vessels and gather fuel. The Moon has a much smaller escape velocity than Earth, so it would take considerably less energy for a launching from the Moon than from the Earth. There would be more energy left for acceleration and deceleration, so the time lapse for a given trip would be smaller. It would not be sufficiently smaller, however, to make the trips practical.

But wait. Because we, and all the life forms we know, live on the surface of a world, we have a natural tendency to find anything else unnatural. In 1974, the American physicist Gerard Kitchen O'Neill (born 1927) suggested the alternative

of artificial settlements for human beings in space. It was not
an altogether new concept and had been used in science fiction
on occasion, but it had never before been put forward in such
careful detail.

O'Neill even suggested two places as bases for humanity;
places that were not on the Moon, but were just as far as the
Moon is from Earth.

Imagine the Moon at zenith, exactly overhead. Trace a line
against the sky due eastwards from the Moon down to the
horizon. Two-thirds of the way along that line, one-third of
the way up from the horizon, at a distance equal to that of the
Moon, is one of those places. Trace another line westward
from the Moon down to the horizon. Two-thirds of the way
along that line, one-third of the way up from the horizon, at a
distance equal to that of the Moon, is another of those places.

Put an object in either place, and it will form an equilateral
triangle with the Moon and Earth. It is 384,400 kilometres
(238,900 miles) from the Moon to the Earth. It is that same
distance from either point to the Moon, or from either point to
the Earth.

What is so special about those places? Back in 1772, the
Italian-French astronomer Joseph-Louis Lagrange (1736–1813)
showed that in those places any small object would remain
essentially stationary with respect to the Moon. As the Moon
moved about the Earth, any small object in either of those
places would also move about the Earth in such a way as to
keep step with the Moon. The competing gravities of Earth
and Moon would keep it where it was.

If the small object were not exactly in the place, it would
wobble back and forth ('librating') about the point. The two
points in space are called Lagrangian points or libration points.

Lagrange discovered five such points altogether, but three of
them are of no practical importance because they represent an
unstable condition. An object would have to remain exactly
at those points to remain at rest with respect to the Moon.
Once pushed out of place, however slightly, the object would
continue to drift away and would never return. The two

points in which an object remains stably in place (except for libration) are those points that form equilateral triangles with the Moon and the Earth. The one that lies towards the eastern horizon is L4 and the one towards the western is L5.

O'Neill suggested that advantage be taken of that gravitational lock and that space settlements be built in the regions around the two libration points, settlements that would become permanent parts of the Earth-Moon system. The settlements themselves could consist of spheres, cylinders, or doughnut-shaped objects that would be large enough to hold 10,000 to 10 million people.

Human beings could live on the inner surface of such objects, which would be set to spinning at a rate that would produce a centrifugal effect that would hold everything and everyone to that inner surface with a force equivalent to Earth's surface gravity. The inner surface could then be designed and contoured into a familiar world. It could be spread with soil, which could be used for agriculture and, eventually, animal husbandry. All the artificial works of man – his buildings and machines – would be there, too.

The material forming the hull of the settlement would be composed of alternations of metal and glass. Sunshine, reflected by large mirrors that would accompany the settlement into orbit, would enter and illuminate the settlement, turning what would otherwise be a cave into a sunlit world. The entry of light could be controlled by louvers over the windows to allow for alternating day and night and to keep the temperature of the settlement equable.

It is from the Sun that the colony would obtain its energy – a copious, easily handled, nonpolluting form of energy.

The larger settlements would have a content of air thick enough to allow a blue sky and to support clouds. Parts of the inner surface of large settlements could be modelled into mountainous territory – full-sized mountains and not just bas-reliefs.

It would be expensive to build such settlements, but the expense would be far less than the world now spends on its various military machines. Since Earth, if it is to survive, will have to practise increasing international cooperation, those military machines will have to wither, and the effort to build settlements in space may well offer us a constructive way to make use of the money and people that are now engaged in war and its preparations.

Besides, the expense of building settlements will decrease as the techniques for the purpose are improved and as the space settlers themselves, in the natural urge to expand their range, take over the building of further settlements.

But where are we to get all the material for the construction of these space settlements? Our groaning planet, sagging under its weight of humanity, with its supply of key resources sputtering and giving out, couldn't possibly afford to give up the colossal quantities of supplies needed for it all. Millions to hundreds of millions of tons of construction material would be needed for each settlement.

Fortunately we have the Moon, a completely dead world with no native life, however simple, whose 'rights' need trouble our sense of ethics.

Lunar material would yield the aluminium, iron, titanium, glass, concrete, and other substances needed for constructing the colony. Lunar soil would be spread over the interior surface. Not only is all that material present in the Moon in huge quantities, but lifting it off the Moon against that body's weak gravity would require only $\frac{1}{20}$ the effort necessary for lifting it off Earth. All the smelting and other chemical work would be done in space.

The lunar material is not perfectly adapted to human needs, to be sure. It is low in the volatile elements carbon, nitrogen, and hydrogen, and these are essential to the functioning of the settlement. Fortunately, the Earth is not short of any of these and can well afford to supply the initial quantities. These would be carefully conserved and recycled, of course, so that replacement supplies would be held to a minimum. Eventually, other sources for volatiles would be exploited – passing comets, for instance.

Dangers and difficulties? Of course.

The possibility of a meteor strike exists, but that is not a very great one. The chance would be far less than that of earthquakes or volcanic eruptions on Earth – which occasionally destroy cities.

Energetic solar radiation is dangerous but would not be a problem in a settlement protected by aluminium, glass, and soil. Cosmic-ray particles offer a more serious problem and the outer hull of the settlement would have to be thick enough to absorb the bulk of them.

Then, too, the centrifugal effect of the cylinder spin would not perfectly duplicate Earth's gravitation. On Earth, the gravitational pull is not perceptibly altered as we rise from the surface. Inside the spinning settlement, however, the centrifugal effect would weaken rapidly as one rose from the inner surface and fall to zero at the axis of rotation of the settlement. We have no way of knowing yet whether such a fluctuating gravitational effect is dangerous to the human body in the long run, but in view of experiences in space so far, we can fairly hope it won't be.

Why should such settlements be built? Human beings are not likely to undertake a vast construction project merely for the fun of doing it. The Great Wall of China was built to hold off the barbarian hordes. The Pyramids of Egypt were built because the religious beliefs of the time made it seem that preserving the body of the monarch was essential to the well-being of the nation. The medieval cathedrals were built for the greater glory of God.

As to space settlements, the motivation may arise out of our declining supplies of petroleum and the difficulty of finding a source of energy large enough, safe enough, and long lasting enough to replace it.

The direct use of sunlight would seem to be one possible solution and that sunlight can be gathered more efficiently in space than on Earth's surface. A solar power station can receive the full range of the Sun's energy, unblocked by atmospheric phenomena. If the station is in Earth's equatorial plane in synchronous orbit, at a height of a little over 35,000 kilometres (22,000 miles), it will be in the Earth's shadow only two per cent of the time over the course of a year.

A number of solar power stations girdling the Earth could solve humanity's energy needs for the indefinite future and could also give Earth's nations a positive reason to cooperate, since building and maintaining the stations would serve as literal lifesavers for each of them alike.

If such solar power stations are understood to be needed and if the effort is made to build them, the space settlements will naturally come into being to house the workers who will serve on the mining stations on the Moon and at the construction sites themseves.

Indeed, beginning with the drive for power stations, space may be put to greater and greater use as observatories, laboratories, and whole factories (much more computerized and automated than they are on the Earth's surface) are lifted into orbit.

With so much of man's industrial and technological activity lifted into space, Earth may return to a more desirable wilderness/park/farm condition. We could restore the beauty of the Earth without losing the material advantages of industry and high technology.

Once the space settlements are established over the next couple of generations as part of a programme for meeting the dire need of Earth's population for energy, there may be a number of ancillary advantages.

As the space settlements increase in number, the room available for human beings would increase, too. Within a century, there could conceivably be room for a billion people on space settlements, and within two centuries there would be more people in space than on Earth.

This prospect does not obviate the need to lower our birthrate in the long run, for if human beings continue to multiply at their present rate, the total mass of flesh and blood will equal the total mass of the Universe in 9,000 years or so.

In fact, it does not obviate the need to lower our birthrate right now, for long before we could put that first billion into space, Earth's population would have increased by twenty-five billion and that would be disastrous. And yet the presence of space settlements would offer a bit of an escape valve; the birthrate need not drop quite as far with space settlements in existence.

In addition to allowing some space for human numbers, the burgeoning clusters of space settlements will lend additional variety to human cultures. Each settlement might well have its own way of life, and some might be quite a distance off the norm. Each settlement might have its own styles in clothing, music, art, literature, sex, family life, religion, and so on. The options for creativity in general and for scientific advance in particular, would be unbounded.

There could even be items of lifestyle unique to the settlements and impossible to duplicate on Earth.

Mountain climbing on the larger settlements would have comforts and pleasures unknown on Earth. As climbers moved higher, the downward pull of the centrifugal effect induced by the settlement's spin would weaken, and it would be easier to climb still farther. Then, too, the air would grow neither thinner nor colder to any substantial degree.

Finally, in carefully enclosed areas on the mountain tops, where the centrifugal effect is particularly low, people could fly by their own muscle power when they were outfitted with plastic wings on light frames, thanks to the thick air and the small downward pull

Space mariners

For the purposes of this book, however, the chief value of the space settlements would be this: They would make possible the exploration of the Solar System – not so much for physical reasons, as for psychological ones.

Consider:

To begin with, space flight is an exotic matter to the people of Earth, something that would take them away from the world on which they live, and on which ancestral life has developed over a period of more than three billion years.

Space flight would, on the other hand, be of the very essence of life to the space settlers. Their worlds would have been populated as a result of space flight; and their labours on the Lunar mining stations and at construction sites would involve space flight as a matter of course.

Tourism would also exist among the multiplying settlements.

Because each settlement would have no perceptible intrinsic gravitational pull of its own, and because all are at roughly the same distance from the Sun, the Earth, and the Moon, travel from one to another would take very little energy. It would be something like coasting on level ice.

Considering the low cost in energy and the fact that various settlements might be each considerably different in culture from others, visitors would have much to be amused by and interested in. It would be quite possible that all space settlers would be space travellers from an early age and the concept would have no terror for them.

Even if the settlers wanted to leave the libration points, or even if they were on the Moon and wanted to leave that world, there would be no need for the strong initial acceleration blast that is required to lift a rocket through the Earth's atmosphere against the Earth's large gravitational pull. That instantly removes the most uncomfortable part of space travel.

Therefore, where Earth people might, on the whole, hesitate to venture into space, and where only a tiny fraction would qualify physically and temperamentally as space explorers, the entire population of the space settlements might be potential explorers.

Then, too, the conditions of space flight represent an extreme changeabout to Earth people. Earth people are accustomed to clinging to the outer surface of a very large world; to a cycling of food, air, and water through so vast a system that one is scarcely aware of it; and to a gravitational intensity that is constant wherever they go.

For space settlers, however, spaceflight introduces changes that are not at all extreme. The settlers live on the inside of a world to begin with; they are aware of and accustomed to a close cycling of food, air, and water; they are accustomed to a variable downward pull.

In short, the space settlers, in undertaking an extended space flight, move from one spaceship to another quite similar, though smaller, spaceship.

All this does not make the space flights to some specific destination necessarily less long or less tedious, but it should enormously lessen the psychological difficulties. A crew of space settlers could undoubtedly endure the restricted quarters of a spaceship over the long flight to Mars and beyond far more stoically and efficiently than a crew of Earth people could.

Again, though, we must ask ourselves for motivation. What would make the space settlers move outwards through the Solar System?

Human curiosity and the desire for knowledge might assure the occasional long-distance flight, but something more would be needed for a mass movement.

That something more is easily seen.

The libration points on either side of the Moon are not very large and could easily be filled. Furthermore, as more and more settlements are built and filled, the drain on Earth's supply of volatile elements would become appreciable and the reluctance of Earth's people to part with them would become pronounced.

It would be useful to search for additional living space and for a better source of volatiles.

The inner Solar System is, as a whole, poor in volatiles. The Moon and Mercury have none, Venus is unapproachable, and Mars, while approachable and possessing volatiles, may not be an ethical source. By the time the space settlements are ready to move outwards, there may be human beings on Martian bases, and the volatiles would, in a way, belong to them.

As I mentioned earlier, comets rich in volatiles wander by

now and then, but this is an intermittent and unreliable source and to depend on them would become ever more risky as the number of settlements multiplies.

It is the asteroid belt that is the nearest appropriate target for expanded living space for the settlements. The many thousands of asteroids might offer even more easily attained construction material than the Moon, and many of them should contain considerable quantities of volatile material.

It may well be that by the twenty-second century, the settlements at the libration points will be recognized as a mere preliminary stage, and it will be the asteroid belt that will be considered the true home of the settlements. They will be farther from Earth and utterly independent of it, but they can remain within radio and television reach of it, of course. There will be endless room out there for the construction of many millions of settlements without crowding.

The outward push might continue even further and belts of settlements might be placed around Jupiter and Saturn at distances large enough to avoid the magnetic fields and the charged particles with which those are filled.

In short, the space settlers will prove the Phoenicians, the Vikings, and the Polynesians of the Space Age, venturing out on a far vaster sea to settle their new lands and islands.

By the twenty-third century, the Solar System may well have been thoroughly explored by human beings with settlements in favoured places throughout. The Sun itself can serve as an adequate energy source if its radiation is properly gathered and focussed, even far out in the vastness of the outer Solar System, and hydrogen fusion reactors should eventually serve as an alternate adequate source.

Stepping stone

This optimistic picture of the total exploration and, so to speak, occupation of the Solar System depends, to a surprising extent, on the use of the Moon as a stepping stone.

Suppose the Moon weren't there in our sky; that it hadn't been formed along with the Earth by some enormously low-probability accident; or that it hadn't been captured late in Earth's life by an equally enormously low-probability accident. Think how that might have affected humanity's technological development.

It was the Moon that first gave human beings the concept of a plurality of worlds. It was the Moon's size and nearness that made it an interesting world and lured us out into space towards what was such a tempting target.

Without the Moon, advancing astronomical techniques might have revealed the planets to be worlds, but would human beings have really tried to develop space travel if the nearest reasonable objects were Venus and Mars, and if flights to the nearest reasonable goal would require a round trip of well over a year?

We need an easy target on which to work out the technology of space flight, and human beings have to be encouraged to strive towards that technology with the bribe of an attainable success.

Of course, human beings might still have sent rockets into space and placed people in orbit around the Earth, even without the presence of the Moon. Such flights have many functions other than that of reaching the Moon. The desire to study Earth as a whole – its resources, its atmosphere and weather pattern, its magnetosphere, the dust and cosmic rays outside the atmosphere, the observation of the rest of the Universe from a position outside the atmosphere, the utilization of solar energy – all would have urged us onwards to rocketry and space exploration.

It might all have been less likely without the Moon beckoning us in our fictional dreams, but given the lapse of additional time it might have taken place. Indeed, without the Moon, we could imagine everything that has taken place so far to have taken place anyway, except for the manned and unmanned flights to and past the Moon. Even the probes to the far-distant planets would have taken place.

But would we then have progressed onwards to space settlements? If such things seem impractical to many 'hardheaded' human beings now, how much more impractical would it seem if all the material for the construction of settlements had to come from Earth itself; if there were no way of using the Moon as a source of raw materials?

And without the space settlements, the true exploration of the Solar System would, in my opinion, be most unlikely.

If, then, it is true that a large Moonlike satellite is a very unlikely possession for an Earthlike habitable planet, and that Earth is the beneficiary of a very rare astronomical accident in this respect, then we must wonder if other civilizations have ever developed space-flight capacity greater than that which we possess right now.

Are other civilizations, one and all, confined to their planet and its immediate environs, and are they capable, at the most, of sending probes to other planets? And is this true, no matter how advanced their technology? It is a tempting thought. It would so neatly explain why the Universe seems so empty, even though half a million and more civilizations may exist in our own Galaxy alone.

It would also offer a sop to our pride. Thanks to our lucky possession of the Moon, it might be that within the next couple of centuries we will develop space-flight capacities far beyond other civilizations that may be far older, and in other respects far more advanced than we. Will we, and no other civilizations, eventually fall heir to the Universe, thanks to the Moon?

It is hard to believe that, perhaps. Surely, given a little more technological development than we have, and given the driving force of the need for energy, a civilization would somehow launch itself into space even without the presence of a Moon. The planet's own resources would be used, at whatever reasonable cost, and the direct flight to the nearest planets made no matter how tedious and difficult. Once that was done the resources of the nearby planet could be used to continue matters.

Perhaps every one of the civilizations would do this — not as easily as we would do it, but perhaps all the better, thanks to the greater intensity of the challenge. Perhaps every civilization develops space flight and explores and settles its planetary system.

In that case, why haven't we heard from them? Why hasn't any civilization come calling?

What would be needed for a visit is not merely the capacity to flit from one planet to another, but from one planetary system to another, and this might represent a completely different order of difficulty.

12 Interstellar flight

The speed of light

The farthest objects we can see in our Solar System are the planet Pluto and its moon, Charon. There are comets that recede to distances far greater than that of Pluto. Perhaps many billions circle the Sun at distances far greater than Pluto at every point. No comet, however, has ever been seen past Pluto's orbit – or past Saturn's, for that matter. The width of Pluto's orbit can therefore be taken as the diameter of the visible Solar System and that comes to 11,800,000,000 kilometres (7,330,000,000 miles).

This is an enormous distance, for the diameter of the visible Solar System is nearly eighty times the distance from the Earth to the Sun. Nevertheless, the distance to even the nearest star, Alpha Centauri, is about 3,500 times that diameter.

If the Solar System were so shrunk that the orbit of Pluto would just fit around the Earth's equator (on which scale the Earth would be 160 kilometres, or 100 miles, from the Sun), Alpha Centauri would be at just the distance of Venus at its closest – and Alpha Centauri is the *nearest* star. Sirius is twice as far away as Alpha Centauri; Procyon 2.5 times as far away; Vega 6 times as far away; Arcturus 9 times as far away; Rigel well over 100 times as far away.

We can look upon these distances in another fashion. Consider the speed of light and of electromagnetic radiation (x-rays, radio waves, and so on). That speed is 299,792.5 kilometres (186,282.4 miles) per second. This is important, since our fastest means of communication is through use of electromagnetic radiation. We know no signals that travel faster.

It takes 1.25 seconds for light (or any similar radiation) to travel from Earth to the Moon. That means that when someone on Earth speaks to an astronaut on the Moon, he cannot possibly get an answer in less than 2.5 seconds, even if the astronaut answers instantly in a mere eyewink on hearing what is said to him.

If we define a 'light-second' as the distance light can travel in one second, then the Moon is 1.25 light-seconds from the Earth.

It takes light 10.93 hours to travel the full width of Pluto's orbit. If we imagined a space settlement at each side of the orbit, with one attempting to establish communication with the other, then the one who spoke first could not expect to get an answer, *under any circumstances we now know of*, in less than 21.86 hours.

Therefore, the diameter of the visible Solar System is equal to 10.93 light-hours; that is, 10.93 times the distance light can travel in one hour.

Using that system, Alpha Centauri, the nearest star, is 4.40 light-years away, or 4.40 times the distance light can travel in one year. If someone on Earth sent a message to a planet circling Alpha Centauri and an answer was sent back the very instant the message was received, the person on Earth would have to wait 8.8 years after sending his message to get an answer.

As for other stars, Sirius is 8.63 light-years away; Procyon, 11.43 light-years; Rigel (still a comparatively close star), 540 light-years away. It would take over 1,000 years to get an answer from a planet circling Rigel.

This might seem irrelevant to the problem of getting to the stars. If light takes 4.40 years to reach Alpha Centauri, need we not merely build up our speed to where it is faster than light and thus outrace the signal and get there in less time than light does?

However, as Albert Einstein (1879–1955) first pointed out in his Special Theory of Relativity in 1905, it is impossible for any object with mass to exceed the speed of light. Einstein set this limit from purely theoretical considerations and it seemed, when it was first suggested, to go against the dictates of 'common sense' (and it seems so to many people even today) – but

it is true just the same. The speed-of-light limit has been verified in innumerable experiments and observations, and there
is no even remotely reasonable ground for doubting it where
the matter and the Universe we know are involved.

The 'common sense' that makes it so difficult to accept the
limitation is based on our experience with everyday phenomena. We notice that if we keep pushing an object, it goes
faster and faster and faster. In fact, Newton's second law of
motion specifically states that this is so and that an equal push
will always result in an equal speed-up regardless of how fast
the object is already moving. It would therefore seem that no
matter how fast we make an object move, we can always make
it go still faster by giving it an additional push. Indeed, careful observation and measurement bear this out under ordinary
circumstances.

But that is because we deal with objects that go only a tiny
fraction of the speed of light, and under such circumstances
Newton's second law does indeed hold as far as we are able to
measure and 'common sense' reigns supreme.

The truth is, though, that if we give an object a push and
make it speed up, and then give it a second push of just the
same size, the amount by which the object speeds up the
second time is *not quite* as high as the first time. Some of the
force of the push goes into increasing the speed, yes; but some
goes into increasing the mass as well.

At ordinary speeds, so little of the force goes into increasing
the mass that that portion is undetectable. As the speed goes
higher and higher, however, a larger and larger fraction of the
force goes into increasing the mass and a smaller and smaller
fraction into increasing the speed, according to a formula
worked out by Einstein. When the speeds are high enough, so
much of the force goes into mass and so little into additional
speed that we begin to notice that Newton's second law and
'common sense' aren't working any more.

It wasn't until the opening of the twentieth century that
scientists knew of any objects that moved fast enough to begin to show the imperfection of the second law. The fast-

moving objects then discovered were subatomic particles, and careful studies of these tiny objects showed that Einstein's equation relating force and speed was exactly right.

By the time the speed of any object gets close to the speed of light, hardly any of the force applied to it goes into additional speed. Almost all of it goes into additional mass. The speeding object becomes much more massive, but hardly any more speedy. In the end, even if you put an infinite amount of force into the speeding object, you can only serve to give it an infinite mass and raise the speed only to the speed of light.

That means that even if you accelerate to maximum speed in an instant by some magic device, it would still take you 4.4 years to reach Alpha Centauri. If you could then decelerate to zero in an instant, turn around, accelerate to maximum speed in an instant, it would still take you 8.8 years to make a round trip.

In actual fact, you would have to accelerate to a very high speed, and that would take a long time if you confine yourself to an acceleration low enough for the human body to endure. It would then take an equally long time to decelerate so that it would be possible to land on a planet in the Alpha Centauri system.

The need to accelerate and decelerate would add about a year to the time it would take to reach a star if we were to travel at the speed of light all the way. Another year would have to be added on the return, and a third year, perhaps, for the time taken in exploration.

Thus, if we count in acceleration, deceleration, and exploration, the time taken to go to any star and return is the speed-of-light round trip plus three years. To travel to Alpha Centauri, explore the system, and return would take 11.8 years – and Alpha Centauri is, I repeat, the *nearest* star.

What's more, as we shall see, there are serious difficulties involved in so long an acceleration and deceleration and in so high a speed, so that it is clear that interstellar travel is a mighty project that might well defeat the most advanced technology.

That is why earlier in the book I suggested that the inability of any civilization to carry through successful interstellar flights is the most logical reason why Earth has been left unvisited. The difficulty of interstellar flight may be such that no extraterrestrial civilization has ever made physical contact with any other, but that each one is confined, now and forever, to its own planetary system.

And that we are confined to ours.

Beyond the speed of light

Let us, however, not give up so quickly. Let us consider that perhaps there is some way of beating the speed-of-light limit. I said earlier that there is 'no even remotely reasonable ground for doubting it [the existence of the speed-of-light limit] where the matter and the Universe we know are involved'. Would it be possible, then, to suspect there might be matter we don't know or aspects to the Universe we don't know?

To begin with, for instance, the speed-of-light limit applies most clearly to objects that – when they are at rest relative to the Universe as a whole – possess mass. This includes all the components of atoms and, therefore, of ourselves, our ships, and our worlds. All these must always travel at less than the speed of light and only infinite force can bring them to the speed of light itself.

That would seem to include everything, but it doesn't. There are some objects that do not have mass, or would not have any if they were at rest relative to the Universe generally. Such objects with 'zero rest-mass' include the photons that are the units of all electromagnetic radiation. It also includes the gravitons that are, in theory at least, the units of the gravitational force. Finally, it includes several different varieties of particles called neutrinos.

All particles with zero rest-mass must, at all times, move through a vacuum at precisely the speed of light, not a hair less, not a hair more. It is because light is made up of photons that go at that speed, that we speak of the 'speed of light'.

If slow-moving particles with mass interact in such a way as to produce a photon, that photon darts off instantly at the speed of light without any perceptible interval during which it accelerates. Again, if a photon is absorbed by some particle with mass, its speed vanishes at once without any perceptible interval of deceleration.

It is sometimes speculated that it might be possible some day to convert all the particles-with-mass in a ship, including those in the crew and passengers, into photons of different types. The photons would then, without the necessity of acceleration and without the expenditure of the energy ordinarily required to bring about that acceleration, move off at the speed of light. Ordinarily, they would move off in all directions, but we might imagine the conversion to take place under conditions that would produce a laser beam of light. Such light would all move off in the same direction, for instance that of Alpha Centauri. Once the photons had arrived at Alpha Centauri, they would be converted back to the original particles – something that would require no deceleration, and none of the energy ordinarily required for such deceleration.

In this way, it would appear that any ship engaged in a round trip to some particular star might save the year ordinarily lost in acceleration and deceleration each way and, what is far more important, would be spared the vast energies required.

There are disadvantages, though. In the first place, it would still mean travel at the speed of light only. Saving two years might be significant, but only for the nearest stars. Allowing one year for exploration, the round trip to Alpha Centauri would take 9.4 years rather than 11.4 years, which is a significant saving; but the round trip to Rigel would take 1,081 years instead of 1,083, which is not.*

Secondly, I am not at all sure that it is possible to divorce speed and energy expenditure as I have so glibly stated. I have a strong suspicion that if we arranged to convert a quantity of

* There is one somewhat more hopeful aspect of such a trip that
  I am omitting now, but that I will come back to later.

matter into photons, we would find that the amount of energy we had to expend to do so would be equal to the amount we would have had to expend to accelerate the matter to near the speed of light in the first place. The same would be true of the conversion back into matter, where we would have to expend as much energy as we would have had to in decelerating the matter from the speed of light. Therefore, it could be that the 'photonic drive' would save us no time to speak of, and no energy either.

Besides, we have no idea how it would be possible to convert matter into photons and then back into matter in such a way as to reproduce all the characteristics of the original matter to the finest details. (Just imagine reproducing a human brain in all its intricacy after it had been dissolved into photons. Some might consider it conceivable, but even those who do can give no hint of the actual method for doing it.)

Then, too, the conversions in either direction would have to be done with very tight simultaneity, for if some conversions into photons are made even a second later than others the photons will be spread out over hundreds of thousands of kilometres, and it might well then be impossible to reconvert them into compact objects.

How might the photons, even if produced in tight simultaneity, be directed in the proper direction, kept from losing order in the long voyage, and then reconverted with equally tight simultaneity?

Granted that 200 years ago the feats of modern-day television might have seemed just as impossible and out of the question, we cannot safely argue that because *some* things that were thought fantastically impossible have proved possible after all, *all* things that seem fantastically impossible *will* be proved possible.

In this book, I have taken the conservative route at all times and accepted nothing without at least some evidence, however slight and tenuous. At the present moment, there is no reason to suspect that a photonic drive can be made practical, and until some evidence to the contrary arrives (and that could be tomorrow, of course) I must say that while I cannot positively rule out a photonic drive, I consider its chances so close to zero that we may reasonably call it that.

Could we avoid the difficulty of conversion and reconversion, and of directing the light beam, by leaving all the particles as particles but somehow removing their mass? A massless ship-and-contents would instantly accelerate to the speed of light and remain at that speed. Once the mass was restored, it would instantly change to its original speed. That seems a much more comfortable situation than conversion into a beam of photons.

Unfortunately, we know of no way of removing mass from any particle, nor is there any indication anywhere that we will ever find a way. And if we did, we would still be travelling only at the speed of light.

So far, all I have suggested brings us to the speed of light, but doesn't pass us beyond it.

In 1962, however, the physicists O. M. P. Bilaniuk, V. K. Deshpande and E. C. G. Sudershan pointed out that Einstein's equations would allow the existence of objects with mass that is expressed by what mathematicians call an imaginary quantity.

Such objects with 'imaginary mass' must always go at speeds faster than that of light if Einstein's equations are to remain valid. For that reason, the American physicist Gerald Feinberg (born 1933) named them *tachyons* from a Greek word meaning *fast*.

An object with imaginary mass would have properties quite different from ordinary mass. For one thing, tachyons have more energy the *slower* they are. If you push a tachyon and thus add energy to it, it goes more and more slowly, until with an infinitely strong push you can make it go as slowly as the speed of light, but never slower than that speed.*

---

* The speed-of-light limit exists for tachyons as well as for particles with ordinary mass (*tardyons*) but in the case of the former, the limit is a floor rather than a ceiling. Particles with zero mass (or *luxons* from a Latin word for *light*) go just at the speed of light,

On the other hand, if you subtract energy by pushing on a tachyon against the direction of its motion or by having it pass through a resisting medium, it goes faster and faster until, when it is at zero energy, it moves with infinite speed relative to the Universe in general.

Suppose, then, we imagine a 'tachyonic drive'. Suppose every subatomic particle making up a ship and its contents is converted into the corresponding tachyons. The ship would take off at once, without acceleration, at many times the speed of light, and reach a distant galaxy in perhaps no more than a few days, at which time everything would be reconverted to the original particles and at once without deceleration, the ship and its contents would be moving at normal velocities.*

Here at last is a way of beating the speed-of-light limit were it not that:

First, we don't really know that tachyons exist. To be sure, they don't violate Einstein's equations, but is that all that is needed for existence? There may be other considerations, outside the equations, that preclude their existence. Some scientists, for instance, hold that tachyons. if they exist, would permit the violation of the law of causality (that cause must precede effect in time) and that this ensures their non-existence. Certainly, no one has detected tachyons so far, and until they are detected, it is going to be hard to argue their real existence, since no aspect of their properties seems to affect our Universe and therefore compel our belief even in the absence of physical detection.†

----

or right at the limit that serves as a boundary, a 'luxon wall' between our own tardyon Universe and the ultrafast tachyon Universe.

* In science fiction stories it has long been customary to get round the speed-of-light barrier by making use of some aspect of the Universe in which the barrier no longer exists. The aspect is called hyperspace or subspace, but whatever the word the imagined properties are those of the tachyonic Universe.

† For twenty-five years, physicists accepted the existence of the neutrino even though it had never been detected, because that

Secondly, even if tachyons exist, we have no idea at all of how to turn ordinary particles into tachyons or how to reverse that process. All the difficulties of the photonic drive would be multiplied in the case of the tachyonic drive, for a mistake in simultaneity of conversion would scatter everything not merely over hundreds of thousands of kilometres but perhaps over hundreds of thousands of light-years.

Finally, even if it could be handled, I still suspect we can't beat the energy requirement; that it would take as much energy to shift matter from one end of the Galaxy to the other by tachyonic drive, as it would by acceleration and deceleration. In fact, the tachyonic drive might take far more energy, since time as well as distance must be defeated.

But we have another possible means of escape. If the qualification 'the matter we know' fails us, what about the 'Universe we know'? As long as the Universe we worked with was that which Newton knew – the Universe of slow movement and small distances – Newton's laws seemed unassailable.

And as long as the Universe we work with is the one Einstein knew – the Universe of low densities and weak gravitations – Einstein's laws seem unassailable. We might, however, go beyond Einstein's Universe as we have gone beyond Newton's. Consider:

When a large star explodes and collapses, the force of the collapse and the mass of the remnant that is collapsing may combine to drive the subatomic particles together into contact – then smash them and collapse indefinitely towards zero volume and infinite density.

The surface gravity of such a collapsing star builds up to the pitch where anything may fall in but nothing may escape again,

---

existence was necessary to explain observed phenomena. Right now, physicists accept the existence of particles called quarks though they have never been detected, because that existence is necessary to explain observed phenomena. There are no observed phenomena that require the existence of tachyons, however, only the manipulation of equations.

so that it is like an endlessly deep 'hole' in space. Since not even light can escape, it is the 'black hole' I mentioned earlier in the book.

Usually one thinks of matter falling into a black hole as being endlessly compressed. There are theories, however, to the effect that if a black hole is rotating (and it is likely that all black holes do), the matter that falls in can squeeze out again somewhere else, like toothpaste blasting out of a fine hole in a stiff tube that is brought under the slow pressure of a steamroller.

The transfer of matter could apparently take place over enormous distances, even millions or billions of light-years, in a trifling period of time. Such transfers can evade the speed-of-light limit because the transfer goes through tunnels or across bridges that do not, strictly speaking, have the time characteristics of our familiar Universe. Indeed, the passageway is sometimes called an Einstein-Rosen bridge because Albert Einstein himself and a co-worker named Rosen suggested a theoretical basis for this in the 1930s.

Could black holes some day make interstellar travel or even intergalactic travel possible? By making proper use of black holes, and assuming them to exist in great numbers, one might enter a black hole at point A, emerge at point B (a long distance away) almost at once, and travel through ordinary space to point C, where one enters another black hole and emerges almost at once at point D, and so on. In this way, any point in the Universe might be reached from any other point in a reasonably short time.

Naturally, one would have to work out a very thorough map of the Universe, with black-hole entrances and exits carefully plotted.

We might speculate that once interstellar travel starts in this fashion, those worlds which happen to be near a black-hole entrance would prosper and grow, and space stations would be established still nearer the entrance.

Those space stations can serve as power stations as well, since the energy radiated by matter falling into a black hole can clearly be enormous. We might even visualize space projects that consist of the moving of dead and useless matter into a black hole to increase the energy ouput (like fuelling a furnace).

In fact, this offers still another explanation for the Universe being full of extraterrestrial civilizations that nevertheless do not visit the Earth. It could be that Earth happens to be in a distant backwater as far as the black-hole networks are concerned. The extraterrestrial civilizations might know all about us, but find us not worth the time and expense of visiting.

Yet the exciting picture of a black-hole-riddled Universe converted into a kind of super-subway-system for interstellar flight has its drawbacks.

In the first place, we don't really know how many black holes there are in the Universe. Outside the centres of the Galaxy and of globular clusters, there might be only half a dozen black holes per galaxy for all we know, and these would be of no use except to a few planetary systems near an opening, none of which might contain a habitable planet.

Second, the suggestion that matter entering a black hole will emerge elsewhere is by no means certain. Many astronomers believe there is nothing to this theory.

Third, even if matter entering a black hole does emerge elsewhere, nothing material can enter a black hole without being thoroughly smashed, right down to a powder of subatomic particles or less, by the incredible tidal effects of the unimaginably intense gravitational field of the black hole. It may be that some advanced technology will learn how to fend off all gravitational effects and keep the spaceship from serving as fuel to the black-hole furnace or from being torn apart by the tides – but at the present moment that seems impossible even in theory.

Looked at in the light of the Universe as it appears to us today, there seems no reasonable hope that the speed-of-light limit will be defeated in any practical way.

We must see what can be done at speeds below that of light.

Time dilatation

One peculiar phenomenon predicted by Einstein's equations (and verified by studies of speeding subatomic particles) is that the rate at which time seems to progress slows with speed. This is called time dilatation.

On a rapidly moving spaceship everything would go more slowly; atomic motions, clocks, the metabolism of human tissue. Because everything on a ship slows down with exact synchronism, people on board such a ship would not be subjectively aware of the change. To them it would simply seem that everything in the outside world had speeded up. (This is analogous to the manner in which one isn't aware of motion in a train moving smoothly forwards at a station; instead the station and the countryside seem to be moving backwards.)

The slowing of time becomes more marked as one moves faster relative to the Universe generally, until by the time a speed of 293,800 kilometres (182,550 miles) per second is reached – 0.98 the speed of light – the rate of time passage is only $\frac{1}{5}$ what it would be if the space vessel were at rest. If the speed of light is approached still more closely, the rate of time's passage continues to drop until, when you are within a kilometre per second of the speed of light, it is nearly zero.

Suppose, then, we are in a spaceship that is accelerating at 1-g. (That is, at a rate that would make us feel pushed against the rear of the ship with the same force that gravitation now pulls us against Earth's surface. At this acceleration we would feel perfectly normal. The back of the ship would seem down, the front up.)

After about a year of this, the ship would be moving at nearly the speed of light and by that time, although everything on board would seem normal to us, the outside world would seem very strange. It would become impossible, really, to watch many of the stars, for the light from stars ahead would shift far into the x-ray range and would be invisible. (In fact, the ship would have to be shielded from their radiation.) The light from the stars behind would shift into the radio-wave range and would be invisible, too.

If the people on board ship measured their speed against the distances they were covering, they would seem to be going at many times the speed of light, for it would take them only a week perhaps to cover the distance between two stars known to be ten light-years apart. If we could watch them from Earth, we would see that it actually took a little over ten years for the ship to cover the distance, but to the time-slowed sense of the people on board, those ten years would seem only a week long.

By making use of time dilation, then, a space vessel would cover enormous distances in times that would seem comparatively short to the people on board. In a length of time that they would experience as sixty years, they would reach the Andromeda Galaxy, which is 2,300,000 light-years away from us.*

Does time dilation solve the problem?

Perhaps not, for there are difficulties. First, to maintain a 1-g acceleration for an extended period of time (or a 1-g deceleration, for that matter) takes enormous quantities of energy, as I indicated earlier.

Suppose we assume the most efficient way of getting energy, interacting equal quantities of matter and antimatter. Such a mixture undergoes mutual annihilation and the total conversion of matter to energy. For a given mass of fuel such a reaction would yield thirty-five times as much energy as hydrogen fusion, and if there is any way of getting more energy than that out of anything, we have no hint of what it might be at present.

---

* If a photonic drive were possible, the rate of time passage to people experiencing the drive would be *zero*. All trips, even to the very edge of the Universe, would seem to take place in an instant. That is why, fast as time dilatation makes astronauts think they are going, they can never beat a ray of light. It may take them only sixty years to reach the Andromeda Galaxy, but when they get there they will find that a light ray would have reached the Andromeda Galaxy before they did.

And yet to accelerate a ton of matter to 0.98 times the speed of light would mean the conversion of about twenty-five tons of mixed matter and antimatter into energy, or the conversion of 100 tons for any round trip, counting two accelerations and two decelerations. If hydrogen fusion were used as the propulsion medium, something like 3,500 tons of hydrogen would have to undergo fusion. In other words, to carry one ton of matter to Alpha Centauri and back – just one ton – would take ten times as much energy as the people of Earth consume right now in one year.

There is the possibility that one need not use fuel to attain the needed energy. The British-American physicist Freeman John Dyson (born 1923) points out that a spaceship whipping around a planet like Jupiter can be enormously accelerated without any ill effects on the astronauts, since every atom of the ship and its contents will be accelerated alike (barring insignificant tidal effect). Indeed, the Jupiter probes, *Pioneer 10* and *Pioneer 11*, were accelerated in this fashion, gaining energy at the expense of the vast pool of gravitational energy of Jupiter and gaining enough speed in this way to be hurled out of the Solar System.

We can imagine spaceships *en route* to some distant star, slipping past a giant planet now and then to gain huge increments of speed – *if* such giant planets happened to be located in convenient places, which doesn't seem at all likely.

Another way of imagining a spaceship's gaining acceleration without fuel is to picture a laser beam shining upon a large 'sail' surrounding the vessel. The laser beam, based on some convenient body in the Solar System, would be trained continually on the sail and would act as one continuous push serving to steadily accelerate the vessel. The laser beam, to remain in being would, of course, consume the vast quantities of energy that the ship was not consuming. (You can't beat the system when it comes to energy.) In addition, it would be more and more difficult to remain on target as the ship moved farther and farther from home base. Finally, the laser beam could not be used to decelerate unless someone at the destination point up ahead could supply an obliging beam in the opposite direction.

Still, if all nonfuel methods failed and a speed-of-light vessel had to use fuel, it might perhaps not have to carry that fuel. It might be able to pick it up as it went along. After all, interstellar space is not truly empty, not an utter vacuum. There are occasional atoms of matter present, mostly hydrogen.

In 1960, the American physicist Robert W. Bussard suggested that this hydrogen might be picked up as the spaceship ploughed through space. The ship would be a kind of 'interstellar ramjet' but since space has much less matter in it than Earth's atmosphere does, the ship would have to sweep up the matter from a far larger volume of space, compress it, and extract energy through hydrogen fusion.

The ship's scoop, in order to be effective, would have to be at least 125 kilometres (80 miles) in diameter when it is passing through those volumes of space where there are clouds of dust and gas, and matter is strewn most thickly. In clear interstellar space, the scoop would have to be as much as 1,400 kilometres (870 miles) in diameter, and in intergalactic space, 140,000 kilometres (87,000 miles) across.

Such scoops, if we imagine them built of even the flimsiest materials, would be prohibitively massive. How would the materials in those scoops be carried out into space; or how much time and effort would it take to assemble them out of matter already in space?

Even if the energy problem is somehow beaten by methods we can't in the least foresee, it remains true that a huge ship travelling very near the speed of light is peculiarly vulnerable. There may be no danger of striking a star, but it may well be that space is fairly full of relatively small bodies from planets down to gravel.

From the viewpoint of the ship, every object in the Universe that happens to be approaching it will be doing so at the speed of light. Such objects will be impossible to avoid, for any conceivable message that heralds their approach (x-rays or anything else) will be travelling only at the speed of light so that the object itself will be hot on the heels of the message. No sooner will a collision warning sound than the collision will take place.

And any massive object colliding with the ship, where the velocity of one relative to the other is that of light, would leave a neat hole in the ship where it entered, where it emerged, and at all intersections in between. The ship might be a Swiss cheese before long.

Even if we discard sizeable particles and assume there is nothing but very thin gas in the volume of gas being passed through – that is enough to make trouble.

As the spaceship accelerates and goes faster and faster, the atoms in interstellar space strike harder and harder, and more and more of them do so per second.

From the standpoint of the spaceship, the oncoming particles will be approaching at very near the speed of light and that will make them, to all intents and purposes, cosmic-ray particles.

Under ordinary conditions, cosmic-ray intensity in space is not particularly deadly. Astronauts have remained in space for more than three months continuously and have survived handily. Moving through interstellar space at the speed of light, however, with every oncoming particle striking with cosmic-ray speed, the ship will be subjected to an intensity of radiation several hundred times that produced by one of our modern nuclear reactors.

Some scientists suspect that this interference by interstellar matter will itself be sufficient to keep space vessels from ever reaching speeds of over $\frac{1}{10}$ that of light – and at that speed the time-dilatation effect is very minor.

Even if all difficulties are overcome, there remains another problem that lies at the very core of relativity. The slowed time sense affects only the astronauts, *not* the people back on the home planet.

Making use of 1-g acceleration and deceleration, and time dilation, to the full, a trip to the star Deneb and back will take astronauts twenty years (even allowing one year in the Deneb system for exploratory purposes). When they return, however, they will find that 200 years have passed on Earth. The longer they travel at this acceleration, the more closely they will creep up to the speed-of-light limit and the more slowly time will pass for them. Thus, the discrepancy between ship-time-passage and Earth-time-passage rapidly increases with distance. A round trip to the other end of the Galaxy will seem to take fifty years to the astronauts, but they will find that some 400,000 years will have passed on Earth. (This would be true to an even greater extreme in the case of the photonic drive.)

One has the feeling that this alone would suffice to make it certain that there would be no great popular demand among the people on Earth (or on any home planet) for investing in stellar exploration by time dilatation. It is difficult enough to get people to deprive themselves of anything now for the sake of having something desirable or even essential come about in thirty years. To invest an enormous effort in something that will return centuries later or hundreds of thousands of years later would not seem to be something we would count on people doing.

Considering, then, the difficulties in energy requirements, in radiation danger, and in time differential, our conservative standards would make it seem that time dilatation is not a practical means, either physically or psychologically, for reaching the stars.

Coasting

Since all methods for travelling near the speed of light or actually beyond it seem to be impractical, we must see what can be done at low speeds.

The advantage there, of course, is that the energy requirements are not exorbitant, nor is the environment of interstellar space then dangerous. The disadvantage rests in the time such a voyage would take.

Suppose a ship were to be accelerated to a speed of 3,000 kilometres (1,860 miles) per second. This would be very fast by ordinary standards since at that speed a ship could travel from the Earth to the Moon in two minutes. Still, it is only $\frac{1}{100}$ the speed of light, so that the time-dilation effect is negligible, and it would take nearly 900 years for the round trip to Alpha Centauri, the nearest star.

Are there any conditions under which a 900-year trip could be endurable?

Suppose the astronauts are immortal. We might decide that in that case, coasting there and back (with comparatively small intervals of acceleration and deceleration) for 900 years would represent a trivial fraction of an endlessly protracted life and would offer no problem.

However, even if the astronauts are immortal, we presume they would have to eat, drink, breathe, and eliminate wastes. That means there would have to be a complex life-support system that would work without fail for nearly 1,000 years. We might imagine it being done, but surely it would be expensive.

Then, too, the astronauts would have to have something to occupy their minds. Comparatively close quarters with no chance for a change in company for nearly 1,000 years could be very difficult to tolerate. It might not be too cynical to suppose that murder and suicide would empty the ship long before the trip is over, for it is much easier to imagine a victory over death than a victory over boredom.

And, of course, we have no real reason to think – at least so far – that we will ever be able to achieve immortality.

But then we can, perhaps, short-circuit some of the difficulties of immortality by changing it to a temporary death followed by a resurrection. In other words, suppose we freeze the astronauts and place them into suspended animation, bringing them back to life only when the destination is in view.

Under such circumstances, the ship can proceed by coasting at low speeds, avoiding the disadvantages of speed-of-light travel, while the astronauts remain as unaware of the passage of time as they would in the case of time dilatation. To them, a voyage of thousands of years would pass in an eyeblink, and when brought back (it is to be presumed), they would not have aged perceptibly. In that way there would be no need for an inordinately reliable life-support system of the usual form; nor would there be the problem of keeping the astronauts occupied and unbored during the long flight.

There are obvious catches, though. The problem of freezing a human being without killing the person and then bringing about a successful revival is not a problem we seem (so far) to have much hope of solving.

Even if we could solve it, there might well be limits as to how long the frozen body could be kept with its spark of life intact. It might not be possible to maintain it throughout the long voyage between the stars. And if we could do that, then we would have to supply the ship with some foolproof system for maintaining the frozen state (a new form of life-support system) and for acting automatically to revive the astronauts at some appropriate moment. A device that can spring to life after some centuries of remaining dormant is not an easy thing to imagine.

The difficulties are enormous, and while we cannot insist that they will never be overcome given enough time, neither can we be certain that the problem will be solved inevitably.

Then, too, while the frozen astronauts are in suspended animation and, as a result, do not age, and are not aware of the passage of time, this is not true for the people back home who sent them off (unless the entire population of the planet undergoes freezing, which we may dismiss as ridiculous). This means that, exactly as in the case of time dilation, the astronauts will return generations later and will experience a profound 'future shock'.

In fact, even in the case of immortality, there would be difficulties. We might assume that if the astronauts are immortal, then the general population of the planet is immortal as well and that after the long trip the astronauts will return to report to the very people who sent them off long ages ago. But life is sure to have followed very different directions on the ship and on the planet, and the two groups of people are certain to be strangers to one another.

It seems quite likely that, under any circumstances so far mentioned, there will be no point to the astronauts' returning home. Exploration of the stars would have to be undertaken on the understanding that the astronauts and the ships will never be seen again. Messages may be sent and received as the centuries and millennia pass, but that would be all.

The question, though, is whether, in that case, human beings would be willing to go into permanent exile. Or whether the home planet would be willing to undergo the expense of sending intelligent beings if all that will come of it are occasional messages received far in the future.

Might it not in that case be more economical, less difficult, and actually more productive, if automatic probes are sent to the stars? The astronomer Ronald N. Bracewell (born 1921) suggested as early as 1960 that other civilizations might well have used this strategy.

We ourselves have taken this tack in connection with the planets. While astronauts have only been able to go as far as the Moon, automatic probes have landed on Mars and Venus and gone past Mercury and Jupiter. We have gained considerable knowledge as a result of these probes, and even if we were of the opinion that human exploration would be preferable, we must admit that where human exploration is impossible, the probes are a reasonable substitute. So far they have produced results that are by no means negligible.

We might, therefore, send stellar probes outwards. The expense would still be enormous, but it would be far less than that involved in sending human beings. We can indulge in greater acceleration, eliminate life-support systems for either living or frozen astronauts, and feel no concern for the psychological welfare of astronauts. Nor need we fear future shock, since there would be no particular reason for an automatic probe to return – and even if it did, it would not matter to it that generations had passed.

We can imagine advanced civilizations sending out very advanced probes, but surely there must come a point of diminishing returns. The more elaborate the probe, the more difficult and uncertain its maintenance would have to be. Over thousands or even millions of years, it is hard to suppose that anything really elaborate would keep working faultlessly. (Surely even the most advanced civilization could not alter the second law of thermodynamics or the uncertainty principle.)

If we go to an extreme, we might imagine a crew of advanced robots as intelligent as human beings, for instance, exploring the Universe as human beings themselves could not. And yet if robots are *that* intelligent, might they not also find themselves vulnerable to the diseases of intelligence – boredom, depression, rage, murder, and suicide?

It might well be necessary, then, to strike some middle ground, and send out probes containing devices elaborate enough to send back as much useful and interesting information as possible, but simple enough to endure through the ages. It seems obvious, though, that this middle ground will result in ships being piloted by devices far less intelligent than human beings.

This, too, may be the answer to the puzzle of why we have not been visited by other civilizations. Perhaps we have been; but not by living organisms. Perhaps probes have passed through our Solar System and have sent messages back on the nature and properties of the Sun and its planets, and, specifically, on the fact that a habitable planet exists in the system. If one has passed recently enough, it might have reported a burgeoning civilization.

Of course, we can't say how often a probe may have passed through, or when the last probe passed, or whether all the probes have belonged to some one particular civilization.* For that matter, it might well be that the probes have outlived

* It is easy to speculate that those UFOs that are not hoaxes or mistakes (assuming there are any that don't fall into one category or the other) are probes, rather than actual extraterrestrial spaceships piloted by living organisms. That is not inconceivable, but, on the other hand, there is no reasonable evidence in favour of this notion. Not yet.

their particular civilization and are sending back messages use-
lessly.

Worlds adrift

A conservative view of the possibilities of interstellar travel
has made it seem that there is no practical way of sending intelligent organisms from star to star and that the best way
would be the use of automatic probes.

So far, however, we have made the assumption that a crew
of astronauts must complete a round-trip voyage to the stars
in the space of a human lifetime – either by going faster than
light, by experiencing time dilations, by possessing extended
lifetimes, or by the use of deep freezing. Every such device
seems impractical.

But then, what if we abandon the assumption and do not require a round-trip in a single lifetime?

Suppose we design a ship that will coast to Alpha Centauri
and take centuries to make the trip. Suppose we do not expect
the astronauts to be immortal or frozen, but to live normal
lifetimes in the normal manner.

Naturally, they will die long before the voyage is completed.
However, astronauts of both sexes are on board, and children
are born to them, and these carry on – and their children do
the same – and their children – for many generations until the
destination is reached.*

An elaborate life-support system is still needed, but the
problem of keeping the astronauts occupied and unbored may
be solved. Having children helps pass the time. Deaths and
births will bring about a steady change in personnel and remove the boredom implicit in a long, long period of the same
old faces. Then, too, youngsters born on the ship will know no

* Lyman Spitzer suggested such generations-long voyages in 1951,
  and the science fiction writer Robert A. Heinlein wrote a
  novelette called 'Universe' on this theme in 1941.

other existence (at least firsthand) and presumably will not be bored.

On the other hand, is any trip worth that? Will there be volunteers who will not only be willing to spend the rest of their lives on board ship, but who will be willing to condemn their children and their children's children to a total life, from birth to death, on board ship? And will the people on Earth be willing to invest in a tremendously expensive project where any benefits to be derived may come only to their descendants 1,000 years hence?

The answer to these questions might be an obvious – no! In fact, the average person might be so horrified at the thought as to feel that merely to ask the question is not quite sane.

Yet that might be only because all through this chapter I have been (without quite saying so) assuming that the space vessels undertaking the long trip to the stars are what we ordinarily think of as 'ships' – like a huge ocean liner, or like the Starship *Enterprise* on the television show *Star Trek*.

As long as we deal with such ships, the objections to a generations-long voyage are difficult, perhaps impossible, to counter – but must we deal with them?

At the end of the previous chapter, I had envisaged a Solar System dotted with space settlements – settlements large enough to constitute worldlike communities in themselves.

Such space settlements would not carry supplies of food and oxygen in the ordinary sense. They would be in functioning ecological balance that could maintain itself indefinitely, given a secure energy source and the replacement of minimal material. Nor would they carry a crew in the ordinary sense of the word. They would be inhabited by tens of thousands, perhaps even by tens of millions, to whom the settlement would be their planet.

The gradual exploration of the Solar System by the settlers and the gradual extension of the range of the settlements to the asteroid belt and beyond would surely weaken the emotional bonds that would hold the settlers to the ancestral Earth and even to the Sun.

The mere fact that to settlers in the asteriod belt and beyond the Sun will be so much farther off and so much smaller will decrease its importance. The fact that it will become harder to use as an energy source as distance increases will encourage the shift to hydrogen fusion, all the more so since there are ample hydrogen supplies in the Solar System beyond Mars. That, in turn, will make the settlements still less dependent on the Sun.

Furthermore, the farther a settlement moves from the Sun, the easier it can develop a speed capable of taking it out of the Solar System altogether.

Eventually, some space colony, seeing no value in circling round and round the Sun for ever, will make use of some advanced propulsion system based on hydrogen fusion to break out of orbit and to carry its structure, its content of soil, water, air, plants, animals, and people out into the unknown.

Why?

Why not?

For the interest of it, perhaps. For seeing what lies beyond the horizon. For the curiosity and drive that has been extending the range of humanity since it came into being, sending bands of people trekking across continents even before civilization began, and now driving them to the Moon and beyond.

There might also be the pressure of mounting population. With ever more space settlements being constructed, there will be increasing pressure on hydrogen supplies, increasing impatience with the growing complexity of intersettlement relationships.

Besides, the trauma of change would be minimal. The settlers would not be leaving home – they would be taking home with them. Except for the fact that the Sun would be shrinking in apparent size and that radio contact with other settlements would become steadily more difficult to maintain (until both Sun and radio contact disappear altogether), there would be no important difference to the people inside the settlement as a result of the changeover from endless circling about the Sun to endless forward movement in the Universe at large.

Nor need the settlers necessarily fear the slow loss of resources through imperfect cycling, or the consumption of their hydrogen fuel. Once a space settlement becomes a free-world, bound to no star, it could find fuel here and there in the Universe.

It might, for instance, work its way through the comet cloud at the very rim of the Solar System, watching for one of the 100 billion comets present there in its native form as a small body of frozen ices. Even as a 'small body', of course, it is a a few kilometres in diameter and would contain enough carbon, hydrogen, nitrogen, and oxygen to supply any loss of volatiles through imperfect cycling for a long time and supply enough hydrogen for fuel for an equally long time. (After all, the free-world will not be accelerating or decelerating very often or very much. For the most part it will be coasting.)

When a comet is found, it may be picked up and placed in tow to serve as a longtime source of material and energy. Given time, and the free-world will have nothing in greater profusion than time, a string of them can be picked up.

And the Universe may not be empty after the comet cloud is left behind. Other stars will have comet clouds surrounding them and there may well be occasional bodies totally independent of stars.

Such a voyage avoids all the difficulties we mentioned earlier. The free-world will be moving slowly so that there will be none of the difficulties of gas resistance and collision, and no energy requirements for extensive accelerations and decelerations. Those on the free-world need be neither immortal nor frozen; they can live normal lives as we do on an extensive world with many people and with Earthlike scenery and a centrifugal effect that produces an Earthlike gravity. Sunlight will have to be artificial, but that can be lived with.

What's more, the free-world will not have been built and invested in by the people of Earth. It will have been built by space settlers, much in the way that American cities were built

by Americans and not by the European nations from which
the Americans or their ancestors may have come. That means
the free-world will not be dependent upon Earth's willingness
to invest.

Nor will the people on the free-world be inhibited by the
thought that their children and their children's children would
pass their entire lives 'on board ship' – that is what they would
have done in any case. Nor will the free-world people be inhibited by the thought that when they return to Earth thousands or millions of years will have passed. It will very likely
never occur to them that they need return to Earth at all.

Perhaps many settlements will convert themselves into free-
worlds. The Solar System, having taken 4.6 billion years to
develop a species intelligent enough to build a technological
civilization capable of constructing space settlements, may
finally 'go to seed'. It may release free-worlds wandering off in
all directions, each carrying its load of humanity in ecological
balance with other forms of life.

It may even be that the home world, Earth, will in the long
run have significance on a cosmic scale only as the source of
the free-worlds. It may continue to serve as a source until such
time as, for one reason or another, its civilization runs down,
falls into decadence, and comes to an end altogether. The
space settlements that do not choose to leave the Solar System
may also shrivel and decay, and only the free-worlds will carry
on a developing and vital humanity.

Eventually, after a lapse of many generations, a particular
free-world may approach a star. It would probably not be an
accident that it does so. Undoubtedly, the free-world's astronomers would study all stars within so many light-years' distance and suggest an approach to one that is particularly
interesting. They might in this way study white dwarfs, neutron
stars, black holes, red giants, Cepheid variables, and so on –
all from a careful, safe distance.

They may also favour approaching stars that are Sunlike in
order to investigate (with some nostalgia, perhaps) the chances
of a civilization in existence there. It could well be that there

will be no impulse whatever to land on an Earthlike planet and to subject themselves to the long forgotten and by now possibly repulsive way of life on the outside of a world. On such an outside, the cycling system would be so large it could not be controlled, the weather would be a tissue of discomfort and vagaries, and the unselected wildlife would be annoying.

If there were small worlds at a distance from the star, at a sufficiently great distance to have icy materials as well as metals and rock – an asteroid belt would be ideal – then it might be time to build a new space settlement from scratch, abandoning the old free-world, which, despite all repairs, might by then be rather battered. (It would also be an opportunity to introduce new designs and technological advances from the hull in.)

There might well be an overwhelming temptation to linger a while, to build settlement after settlement in the new asteroid belt.

The advantages to this are obvious. During all the long years the free-world has wandered through space, it will have had to maintain a rigid population control. Now there will be a chance to expand population with wild abandon.

Again, through all the long years, the free-world, while much larger than what we would ordinarily think of as a space vessel, would be small enough to make it necessary to enforce a certain uniformity of culture and way of life. The building of numerous space settlements over a period of centuries in an asteroid belt would allow the establishment of widely different cultures.

And, of course, the new space settlements would eventually go to seed and move outward as a new generation of free-worlds.

We might almost imagine civilizations as existing in two alternating forms: a motile, population-controlled form as free-worlds drifting through space; and a sessile, population-expanding form as space settlements about a star.

Each free-world as it drifts through space eventually loses all contact with its home base, with space settlements, with other free-worlds. It becomes a lonely, self-contained culture that develops a literature of its own, as well as art forms, philosophy, science, and customs, with some Earth culture as a distant base, of course. Every other free-world does the same and no one of them is likely to duplicate the culture of another at all closely. And with each settlement in a new Solar System and eventual breakout, a new explosion of difference would result.

Such cultural variations could produce an infinite richness to humanity as a whole, a richness that could only be faintly hinted at if humanity were confined to the Solar System for ever.

Different free-world cultures might have a chance to interact when the paths of two of them intersected.

Each would be detected by the other from a long distance, we might imagine, and the approach would be a time of great excitement on each. The meeting would surely involve a ritual of incomparable importance; there would be no flash-by with a hail-and-farewell.*

Each, after all, would have compiled its own records, which it could now make available to the other. There would be descriptions by each of sectors of space never visited by the other. New scientific theories and novel interpretations of old ones would be expounded. Differing philosophies and ways of life would be discussed. Literature, works of art, material artifacts, and technological devices would be exchanged.

There would also be the opportunity for a cross-flow of genes. Any exchange of population (either temporary or permanent) might be the major accomplishment of any such meeting. Such an exchange might improve the biological vigour of both populations.

To be sure, in the course of the long separation, enough

* It is conceivable that particular free-worlds might be isolationist, fearful or suspicious of other free-worlds, and might choose to veer away from the approach of another. Surely this would not happen often, however. I have better hopes of the curiosity of intelligent creatures.

mutation might have taken place to make the two populations mutually infertile. They might have evolved into separate species, but even so, intellectual cross-fertilization may be possible (provided always that the inevitable language difficulty is overcome, for even if two free-worlds had begun with the same language, these would have developed separately into widely different dialects).

In this way, humanity would become no longer a creature of Earth or of the Solar System, but would belong to the whole Universe, drifting outwards, ever outwards, forming a variety of related species, until such time as the Universe finally came to an enormously slow end and, through one route or another, could no longer support life anywhere within itself.

But what about the extraterrestrial intelligences? Assuming that they do not make use of any dream technologies we cannot even imagine at present, they too may have followed a development that makes the free-worlds a practical way (perhaps the only practical way) of sending living organisms through interstellar space.

Free-worlds may thus arise from thousands of different planetary sources, and some of them may have been moving through space, into and out of the asteroid belts of this star and that, for billions of years.

It may be that if extraterrestrial civilizations have visited us, it has been in the form of free-worlds. And if so, it may be that they have not visited Earth (in which their interest might be limited), but our asteroid belt.

It may be that when our space settlements move out into the asteroid belt we will find ourselves preempted; or perhaps find evidence that free-worlds have been there in the past and have long since gone.*

Or it may be that free-worlds, on principle, avoid Sunlike

* Those with a more romantic imagination might even suggest that there was an intact planet in the orbit between those of Mars and Jupiter; that a free-world dismantled it in order to build numerous space settlements over a long period of time; and that the asteroid belt is the remnant they left behind.

stars with habitable planets. After all, for free-world purposes almost any star would do. A star might be a short-lived giant, but the free-world can stay far enough away to avoid the radiation and might not need more than, say, a century or two to build new starships out of what planetary material is available at such a distance. Even the least long-lived star would last many times that period. Or (much more likely) a star might be pigmyish and cool, but the free-world would not need it for energy, just for the planetary bodies circling it.

If many civilizations adopt that technique, it may well be that some human free-world, dropping down towards some planetary system, will find it already preempted by other free-worlds that are nonhuman.

Surely by that point in history, it will be understood that it is the nature of the mind that makes individuals kin, and that the differences in shape, form, and manner are altogether trivial.

It may be that as the human free-worlds start moving outwards, they will find themselves part of a vast brotherhood of intelligence; part of the complex of innumerable routes by which the Universe has evolved in order to become capable of understanding itself.

And it may be that in combination, humanity and all the extraterrestrial civilizations can advance farther and faster than any one of them could alone. If there is any chance of defeating what we now see as the laws of nature and of bending the entire Universe to the will of the intelligences it has given rise to, then it will be in combined effort that the greatest chance of success will arise.

13 Messages

Sending

We have concluded, then, that there may well be over 500,000 civilizations in the Galaxy, but that the only way any of them are likely to emerge from their planetary systems is by inter-stellar probes or in the form of free-worlds.

There is nothing compelling about either emergence. The vast majority of civilizations, conceivably all of them, may simply remain in their own planetary systems. Any interstellar probes that are sent out may be devices not designed to land on habitable planets but to confine themselves to observations and reports from space. And free-worlds that may come our way might be more interested in material and energy with which to maintain themselves than in involvement with a sedentary civilization.

In this way, we can rationalize the apparent paradox that while the Galaxy may be rich in civilizations we remain un-aware of them.

But what ought we to do in that case?

The simplest answer and the one that involves the least trouble is to do nothing at all. If extraterrestrial civilizations can't or won't reach us, we could just go about our own busi-ness. Certainly we have enough troubles of our own to occupy us.

The second possibility is to send out some sort of message in order to make contact. Even if an extraterrestrial civilization can't reach us, or we them, we can perhaps establish com-munication across space; even if it is only the message: 'We are here. Are you there?'

This is such a normal impulse that back in the nineteenth century, when people were still speculating concerning life on other worlds in the Solar System and almost taking it for granted that there would be civilizations even on the Moon, there were suggestions for methods of communication.

The German mathematician Karl Friedrich Gauss (1777–1855) once suggested that lanes of forest be planted on the steppes of central Asia in the form of a gigantic right-angled triangle with squares on each side. Within the triangle and squares, grain would be planted to darken the shapes with a uniform colour. A civilization on the Moon or Mars, for instance, closely studying the surface of the Earth, might see this clear display of the Pythagorean theorem and would conclude at once that there was intelligence on Earth.

The Austrian astronomer Joseph Johann von Littrow (1781–1840) suggested instead that canals be dug, and that kerosene arranged in mathematical forms be floated on the water and set on fire at night. Again, mathematical symbols would be seen from other worlds.

The French inventor Charles Cros (1842–1888) suggested something more flexible – a vast mirror that could be used to reflect light towards Mars. It could then be so manipulated as to send the equivalent of Morse code and actual messages could, in this way, be sent (though they might not necessarily be interpreted, of course).

Interest in establishing communication with extraterrestrial civilizations mounted to the point where, in 1900, a prize of 100,000 francs was offered in Paris to the first person to carry through this task successfully. Communication with Mars was excluded, however. That was thought to be too easy a feat to be worth the money.

All such nineteenth-century suggestions are useless, of course, since there are no intelligent beings on the Moon, Venus, or Mars, and it is doubtful whether the unsophisticated techniques suggested could reach farther (if, indeed, that far).

Besides, in the twentieth century we have, ironically enough, sent out even more spectacular messages with no special effort on our part.

The invention of the electric light and the gradually increasing illumination of our cities and highways has steadily intensified the glitter of Earth's surface at night, at least over the land areas that are industrialised and urbanized. Astronomers on Mars, puzzling over the light emerging in steadily increasing intensity from Earth's dark side would be sure to come to the conclusion that a civilization existed on Earth – if there were astronomers on Mars.

The nineteenth-century suggestions made use of light, since that was the most easily manipulable radiation known to cross the vacuum of space at that time. Around the turn of the century, however, radio waves were discovered (like light waves, but a million times longer) and put to use. By 1900, the Yugoslavian-American inventor Nikola Tesla (1856–1943) was already suggesting that radio waves be used to send messages to other worlds.

No deliberate attempt of the kind was made, but it didn't have to be. With the passing decades radio waves were generated by human beings with ever increasing intensity. Those that could penetrate the upper layers of Earth's atmosphere did so, and as a result there is a sphere of radio-wave radiation swelling out from Earth in every direction.

Again, astronomers on Mars, if they were aware of this radiation and if they noted that it was growing steadily stronger, would be forced to come to the conclusion that there was a civilization on Earth.

By the second half of the twentieth century, however, it was quite clear that extraterrestrial civilizations did not exist in the Solar System and that if we were to send messages it would have to be to the stars.

This introduced formidable complications. In the Solar System, we at least know where we might aim our messages – at Mars, at Venus, and so on. There is, on the other hand, no way of knowing which star it would be best to aim at.

Furthermore, radiation aimed at the stars would have to be very energetic if it were to maintain sufficient intensity, in view of inevitable dispersion over the light-years, for it to be picked up at even the distance of the nearest stars.

We are, as I have already said, sending out radio-wave radiation to the stars quite involuntarily. The radio waves that have leaked through the upper layers of our atmosphere have expanded now into a vast ball dozens of light-years in diameter. The outer fringes have passed by many stars already, and although the intensity is excessively minute, it could conceivably be picked up.

However, signals so excessively weak might not seem to the distant astronomers to be incontrovertible proof of a civilization existing somewhere in the neighbourhood of our Sun. Even if the astronomers came to the conclusion the civilization existed, the complicated mix of signals would be impossible to sort out and make sense of.

A deliberately emitted beam of radiation could be designed to contain a great deal of information and could be made strong enough to remove all doubt even if its content could not be interpreted.

The trouble is that we do not at the moment want to dispose of the energy to spray messages out into space, especially since we aren't sure of any specific target, and cannot honestly have much hope of an answer until, at best, many years have passed.

Is there something we can do that will cost less in terms of energy?

We might send a material message, something we can cast arbitrarily into space at little or no cost. To be sure, a material message would be harder to aim than a beam of radiation, and the material message might take many thousands of times longer to get to any specific destination, but at least it would be well within our present capacities.

And the fact is that we *have* sent a message.

On 3 March 1972, the Jupiter probe, *Pioneer 10*, was launched. It passed by Jupiter in December 1973, making its closest approach on 3 December, and very successfully sent back photographs and other data that enormously increased our knowledge of that giant planet.

If that were all – if, after having passed Jupiter, *Pioneer 10* had vanished, or exploded, or simply gone dead – it would have proved worthy of the time, effort, and money expended on it. Anything it could do beyond the Jupiter mission was, in a way, an added bonus. Adding a message to it, therefore, would cost virtually nothing.

Pioneer 10 does carry a message, one that was added at the last minute as a matter of sheer bravado.

The message is a gold-anodized aluminium plate, six inches by nine inches, which is attached to the antenna support struts of *Pioneer 10*.

Etched on to the plaque is informational matter that was decided on by the American astronomers Carl Sagan and Frank Donald Drake. Most of the information would be completely over the heads of all but a very few human beings. It involves details concerning the hydrogen atom, and that information is expressed in binary numbers. It locates the Earth relative to nearby pulsars, giving the periods of the pulsars in binary numbers. Since pulsars are in a particular place only at particular times, and since their rate of rotation slows so that they will have the given rate for only a period of time, this information tells exactly where the Earth has been relative to the rest of the Galaxy at a particular time in cosmic history.

There is also a small diagram of the planets of the Solar System and an indication of *Pioneer 10* itself and the path it took in going through the Solar System.

The most noticeable item on the plaque, though, is a diagrammatic representation of *Pioneer 10* and in front of it, to scale, an unclothed man and woman (drawn by Linda Salzman Sagan, Carl's wife). The man's arm is lifted in what (it is hoped) will be interpreted as a gesture of peace.

If an intelligent species should happen to pick up the message, will it be understood? Since it is almost as certain as anything can be that it will be picked up only by some species in a spaceship or a free-world, we can suppose that species will have developed a technology that will possess advanced scientific concepts. They should, therefore, certainly grasp the meaning of the purely scientific symbols. Sagan points out, however, that it is the drawing of the human beings that may puzzle them, since the pictures may be like no form of life they have ever encountered. They may not even interpret the markings as representing a life form.

They will also have *Pioneer 10* itself to study and, in some ways, that may tell them more about Earth and its inhabitants than the plaque will.

But where is *Pioneer 10* taking the plaque? *Pioneer 10*, as it skittered around Jupiter, gained energy from Jupiter's vast gravitational fields, and by 1984 it will coast past Pluto's boundary at a speed of 11 kilometres (7 miles) per second. That will be enough to carry it indefinitely away from the Sun, wandering on for billions of years unless it strikes an object large enough to destroy it.

It will take *Pioneer 10* about 80,000 years to recede from us to a distance equal to that of Alpha Centauri. It will not be anywhere near Alpha Centauri at that time, however, for it is not going in that direction.

Pioneer 10 was not aimed with any star in mind, after all. It was aimed at Jupiter in such a way as to give us maximum information about that planet, and whatever direction it took up thereafter, on leaving the Solar System – that was it.

As it happens, *Pioneer 10* will be following a path that will not come close enough to enter the planetary system of any star we can see for at least ten billion years. Of course it may through sheer accident skim by a free-world some time in its long journey. The chances of even that must surely be exceedingly small, however, and no one seriously expects that *Pioneer 10* will come within the purview of any intelligent species at any time in its long journey.

In that case, why should we have bothered?

In the first place, it was a very small bother. And in the second place, it just might be picked up at some time, and even if those who pick it up are much too far away from us to do anything about it, or if it is picked up at a time long after humanity is extinct, we would nevertheless have made some mark on the Universe.

We would have left behind evidence that once there was an

intelligent species on our small world that could manage to put together enough expertise to hurl an object out of our Solar System. There is such a thing as pride!

Finally, we can multiply our chances by sending out more than one message. An identical plaque was placed on *Pioneer 11*, which will eventually leave the Solar System on a track different from that of *Pioneer 10*.

And in 1977, probes were launched on which were included numerous photographs showing widely mixed aspects of life on Earth, together with a recording containing enormously varied sounds produced on Earth.

Receiving

Obviously, it will be some time before we are in a position to send out messages that are more than passive cartoons, aimed virtually at random.

Furthermore, there is some opposition to the thought of sending out messages at all. The nub of that opposition rests with the question: 'Why attract attention?'

Suppose we do announce our presence. Are we not simply inviting civilizations advanced beyond ours, which have hitherto not been aware of our presence, to make for us at full speed and to arrive with the intention of taking over our world, of reducing us to slavery, or of wiping us out?

The chances seem to me to be strongly against that. I have explained earlier in the book why I consider it very likely that civilizations that have advanced beyond our own level of technology will be peaceful. Even if not peaceful, civilizations are very likely confined to their own planetary systems. In the very unlikely case that a civilization is warlike and is also roaming freely through space, it has probably examined all stars and is aware of our presence. Finally, even if it has unaccountably missed us, we have already given ourselves away by our radio broadcasts.

For all these reasons, it makes no difference whether we

signal or not, and yet it is hard to answer the unreasoning fears that assume the very worst combination of possibilities. Suppose there *are* civilizations out there as vicious and warlike as we ourselves are at our worst, who *can* move through space freely, who *are* looking for new prey, and who have until now been unaware of us. Shouldn't we lie low and keep absolutely quiet?

Accepting that argument, should we not, for our own safety, find out as much as we can about these hypothetical monsters even while we are lying low? Shouldn't we want to know where the danger is, how bad it might be, how best we might defend ourselves, or (if that is impossible) how best we might more effectively hide?

In other words, abandoning any attempt to send messages (at which we are ineffective, in any case) ought we not to make every attempt to *receive* messages? If we do receive a message and decipher it and decide we don't like what we hear, there is, after all, no reason why we would have to answer it.

Would we, however, know we had come across a signal if we detected it? What ought we to look for?

We might take the optimistic attitude that though we can't predict what the signals would be, we would recognize them if they were there. The detection of what seemed to be a network of Martian canals was a complete surprise, but was quickly taken as an indication of a high civilization.

We know now, though, that if life signals are obtained from anywhere it will have to be from the planetary systems of other stars (or possibly from automatic probes or free-worlds in interstellar space). The likelihood is that any signals we do get will come from many light-years away, and the question is whether it is reasonable to suppose that signals energetic enough to make themselves felt across such distances could be sent out.

It might be that we should not judge all civilizations by our own. What seems a high energy level to ourselves might not seem high at all to more advanced civilizations. In 1964, the Soviet astronomer N. S. Kardashev suggested that civilizations might exist at three levels. Level I is Earthlike and can dispose of energy intensities of the kind available through the burning of fossil fuels. Level II could tap the entire energy of its star, thus disposing of energy intensities 100 trillion times that of Level I. Level III could tap the entire energy of the galaxy of which it is a part, thus disposing of energy intensities 100 billion times that of Level II.

A signal from a Level-II civilization could easily have enough energy content to be detectable from any part of the galaxy of which it is part. A signal from a Level-III civilization could easily have enough energy content to be detectable anywhere in the Universe.

We might dismiss this at once by saying that we detect no signals anywhere but, in the first place, we are not really listening. In the second place, even if the signals forced themselves upon our consciousness, would we recognize them for what they are?

In 1963, for instance, the Dutch-American astronomer Maarten Schmidt (born 1929) discovered quasars, extraordinarily bright and distant objects that show irregular variations in brightness. In 1968, the British astronomer Anthony Hewish (born 1924) announced the discovery of pulsars, which send out regular pulses of radiation at very short but very slowly lengthening intervals. Beginning in 1971, certain intense x-ray streams that varied irregularly in intensity were ascribed to black holes.

Could it be that these objects represent the signal beacons of Level-II or Level-III civilizations? To be sure, the variations in intensity seem to be quite irregular in the case of quasars and black holes, and quite regular in the case of pulsars, and in either case don't seem to have the kind of information that would be of intelligent origin – but may that be merely the result of our own inadequate understanding?

Perhaps! From the conservative position of this book, however, it is an extremely unlikely *perhaps*. We can only say that thus far there is no large-scale phenomenon in the Universe, involving the kind of energy output characteristic in intensity of stars or galaxies, where there is any evidence whatever of intelligent information content. Until such evidence arrives, we must delay a decision.

Of course, a signal might not be a deliberate beacon but the entirely involuntary accompaniment of a civilization's activities. We are illuminating our cities and highways only for the convenience and safety of human beings, but it turns out to be a signal to any extraterrestrial civilizations that are close enough and attentive enough to note it.

If the Martian canals really existed, they would do so only to supply the Martian civilization with badly needed water for irrigation – but their existence would have signalled us.

In the same way, a more advanced civilization may do something sufficiently enormous to make itself felt at stellar distances.

Freeman J. Dyson suggested that if human beings began to exploit and explore space, they might wish to expand their numbers to the utmost that can be sustained by the Sun's energy. At the present moment, the Earth stops only a tiny fraction of sunlight, and almost all the solar radiational energy slips past the cool bodies of the Solar System to streak into and through interstellar space. Human beings might therefore eventually break up the various outer bodies of the Solar System to make up a group of free-worlds that will be placed in a spherical shell about the Sun at the distance of the inner edge of the asteroid belt.

All the Sun's energy would be absorbed and utilized by one or another of the free-worlds. The energy would, of course, be reradiated into space from the dark side of each of the free-worlds, but only as infrared radiation. Viewed from another star, then, the Sun's radiation would seem to change its character from one in which a major portion was emitted as visible light to one in which almost all was emitted as infrared. The changeover would take perhaps a couple of centuries, the barest instant of astronomical time.

If, then, from our own Earth we should see some other star, which has been shining steadily as far as our records tell us, suddenly begin to lose brightness and after a while blink out, we can be reasonably sure we have seen intelligence at work.

Well, perhaps – but we haven't seen anything of the sort as yet.

We must come to the conclusion, then, that (1) we are hopelessly inept at detecting signals and might as well not bother; or that (2) no signals are being sent out and that we might as well not bother; or that (3) signals are being sent out but at much less than heroic energy content, and as a result of much less than heroic civilizational activity, and that in order to detect them we will have to make a considerable effort.

Clearly, we cannot accept the first or second conclusions until we have made an honest attempt at the third.

Then let us consider signals of low-energy content (but high-energy enough to detect) and see what they might be like.

They would have to consist of some phenomenon that could cross vast reaches of space, and these can be divided into three classes: (1) large objects such as plaques, probes, and free-worlds; (2) subatomic particles with mass; (3) subatomic particles without mass.

The large objects we can eliminate at once. They move slowly and are extremely inefficient as carriers of information.

The subatomic particles with mass can be divided into two subclasses, those without electric charge and those with electric charge. Subatomic particles with mass but without electric charge generally move slowly and can be eliminated as impractical for that reason.

Subatomic particles with both mass and electric charge *can* move quickly because they are accelerated by the electromagnetic fields associated with stars and with galaxies as a whole. Therefore, in crossing interstellar and intergalactic spaces, they achieve very nearly the speed of light and, in consequence, enormous energies.

Such subatomic particles do indeed occur everywhere and they are constantly and eternally bombarding the Earth. We call them cosmic rays.

The difficulty here, though, is that the mere fact that these particles are accelerated by electromagnetic fields means that they experience an attraction or a repulsion and that, in either case, their paths curve. As the particles gain increasing energy, their paths curve more and more slightly, but over vast distances even the slightest curve becomes important. What's more, a beam of particles is gradually dispersed, since those with more energy are curved less than those with less energy.

The cosmic-ray particles bombard us from all sides, but because of their past experiences with electromagnetic fields, there is no way of telling from the direction of their arrival where they came from. Nor can we tell whether a particular group that arrives together left together. For a signal to be of any use, it has to come in a straight line and be neither dispersed nor distorted, and that eliminates all subatomic particles with mass.

We are now left only with the subatomic particles without mass, and there are only three known general classes of such particles: * neutrinos, gravitons, and photons.

Being massless, all these particles travel at the speed of light and there can be no faster messengers. That is one point in their favour.

Moreover, no massless particle carries an electric charge, so none is affected by electromagnetic fields. They *are* affected by gravitational fields, but detectably so only in regions where such fields are very intense. Even there, beams of massless particles would bend in unison and would not be dispersed. Since the intensity of the gravitational field in space is negligible almost everywhere, all massless particles reach us in essentially a straight line and essentially undispersed and undistorted, even though their origins are billions of light-years away. This is a second point in their favour.

In the case of neutrinos, however, reception is extremely

* If there are other classes that are unknown, then we would not, in any case, detect any messages sent by way of them.

difficult, since neutrinos scarcely interact with matter at all. A stream of neutrinos could pass through many light-years of solid lead without more than a small fraction of them having been absorbed.

To be sure, a *very* small fraction can be absorbed even in relatively small samples of matter, and so many neutrinos can very easily be produced that such a very small fraction might suffice to carry a message.

However, the type of nuclear reactions that go on in the interior of stars produces neutrinos. In a Sunlike star, vast numbers of neutrinos are produced in this fashion.* A civilization is not likely to produce more than an insignificant fraction of the neutrinos their own star will be producing, so that there will be the danger that whatever message the civilization sends out will be swamped by the much greater volume of neutrinos the star is emitting. (It is a general rule, perhaps, that the medium you use for your message should be easily distinguished from the background. You don't whisper a message across a room in a boiler factory.)

There is a possible way out of this. While the fusion reactions involving hydrogen nuclei at the centre of the stars produce neutrinos, the fission reactions involving the breakup of massive nuclei such as those of uranium and thorium produce related particles called antineutrinos.

Antineutrinos are also massless and chargeless but are, so to speak, mirror images of neutrinos. When absorbed by matter, antineutrinos produce different results than neutrinos do, and if a civilization is careful to allow a stream of antineutrinos to be the message carrier, it could be read even in the presence of a vast flood of neutrinos.

Nevertheless, the difficulty of intercepting such particles is

* I feel by no means as certain in making this statement as I would have been a few years ago. Over the last few years there have been attempts to detect the neutrinos produced by the Sun and far fewer have been detected than should have been detected. Astronomers have not yet made up their minds as to the significance of this.

such that no civilization would use this method if something better were available.

Gravitons, which are the particles of the gravitational field, are certainly not better. Gravitons carry so minute a quantity of energy that they are even more difficult to detect than neutrinos. What's more, they are far more difficult to produce than neutrinos. To produce even barely detectable gravitational radiation, using the technology currently at our disposal, huge masses must be made to accelerate – through rotation, revolution, pulsation, collapse, and so on – in some pattern that will serve as a code. We can fantasize a civilization so advanced that it can make a giant star pulse in Morse code, but even that advanced a civilization wouldn't bother if something simpler were available.

That leaves the last category of communication systems – photons.

Photons

All electromagnetic radiation is made up of photons, and these come in a wide variety of energies,* from the extremely energetic photons of the shortest-wave gamma rays to the extremely unenergetic longest-wave radio waves. If we consider any band of radiation in which energy doubles as we pass from one end of the band to the other (or the wavelength doubles in the other direction) then that is one octave. There are scores of octaves making up the full stretch of electromagnetic radiation, and visible light makes up a single octave somewhere in the middle.

All objects that are not at absolute zero in temperature radiate photons over a wide range of energies. There are relatively few at either end of the range, and a peak somewhere in the middle. The peak represents photons of a certain energy,

---

* Or wavelengths. The longer the wavelength, the lower the energy; the shorter the wavelength, the higher the energy.

and as the temperature rises, the peak is located at higher and higher energies.

For very frigid objects near absolute zero, the peak radiation is far in the radio-wave region. For objects at room temperature, like ourselves, for instance, the peak is in the long-wave infrared. For cool stars, it is in the short-wave infrared, though enough photons of visible light are radiated to give the stars a red colour. For Sunlike stars, the peak is in the visible-light region. For very hot stars, it is in the ultraviolet, although enough photons of visible light are produced to give the star a blue-white appearance.

Most of the range of electromagnetic radiation cannot penetrate our atmosphere, but visible light can, and most organisms have evolved sense organs that can respond to these photons. In short, we can see.

On Earth, at least, we have the aid of our other senses, but for any object beyond our atmosphere, the only information we have ever received (until very recently) is through the visible-light photons that have reached us from those objects.

It is natural, therefore, that we would think of signals from outer space in terms of visible light. We *see* the Martian 'canals' and extraterrestrials watching Earth would *see* any markings we deliberately drew on the planetary surface, or the lights of our nighttime illumination.

Signalling by light represents a vast advance over signalling by neutrinos or gravitons. Light is easily produced and easily received. We can imagine some civilization setting up an exceedingly intense beam of light, and flicking it on and off in some way that would make it instantly recognizable as the product of intelligence. For instance, if we represent each flick as *, we might receive, over and over again, **_***_ ***** _ ******* _ ************ _ ************* _ ***************** _ We would recognize that at once as the first members of the series of prime numbers and could not doubt that we were dealing with a signal of intelligent origin.

There are difficulties, though. A light beam intense enough

to be seen at interstellar distances would require vast energies, and even then the light beam would be completely drowned out by the light of the star that the planet circles.

A Level-II civilization might conceivably know of ways to make a star bright and dim in such a way as to make a signal of undoubted intelligent origin, and a Level-III civilization might make a whole group of stars do so. This, however, is pure speculation. Nothing like it has ever been observed and it would certainly be unnecessary to make use of so heroic a signalling device if we can find something simpler.

For instance, what if the signal beam were a kind of light that was not produced in nature? This suggestion might have seemed silly prior to 1960, but in that year the laser was developed by the American physicist Theodore Harold Maiman (born 1927), and within a year it was suggested as a possible carrier for interstellar messages.

All light produced in ordinary fashion is 'incoherent'. It comes in a wide band of photon energies, and the different photons are generally heading different ways. A beam of such light quickly spreads out no matter how we try to focus it; and to keep it intense enough to be detectable at interstellar distances requires almost stellar energies.

In a laser, though, certain atoms are lifted to a high energy level and are allowed to lose this energy under conditions that produce 'coherent' light – light that is composed of photons that are all of equal energies and are all moving in the same direction. A laser beam scarcely spreads out at all, so that for a given energy it can remain intense enough to be detected at far greater distances than a beam of ordinary light. What's more, a beam of laser light can be easily identified spectroscopically, and merely through its existence is satisfactory indication of intelligent origin.

With laser light we come closer to a practical signalling device than anything yet mentioned, but even a laser signal originating from some planet would, at great distances, be drowned out by the general light of the star the planet circles.

One possibility that has been suggested is this:

The spectra of Suntype stars have numerous dark lines representing missing photons – photons that have been preferentially absorbed by specific atoms in the stars' atmospheres. Suppose a planetary civilization sends out a strong laser beam at the precise energy level of one of the more prominent dark lines of the star's spectrum. That would brighten that dark line.

If we studied the spectrum of a star and discovered that it was missing one of the dark lines characteristic of a certain group of atoms in the star's atmosphere, but that other dark lines also characteristic of that group were present, we would have to conclude that the missing energy level had been supplied by artificial means. That would mean the presence of a civilization.

Nothing like that has been observed – but before feeling depressed over that, let us see if perchance there are still simpler ways of signalling. After all, no civilization would be expected to use the harder method when a simpler is available.

Microwaves

Early in the nineteenth century, electromagnetic radiation outside the range of visible light was first discovered. In 1800, William Herschel discovered the infrared range of sunlight by the manner in which a thermometer was affected beyond the red limit of the range of visible light. In 1801, the German physicist Johann Wilhelm Ritter (1776–1810) discovered the ultraviolet range of sunlight by the manner in which chemical reactions were brought about beyond the violet limit of the range of visible light.

These discoveries did not affect astronomy very much, however. Most of the range of ultraviolet and infrared could not penetrate the atmosphere, so that little of it reached us from the Sun and the stars.

Beginning in 1864, Maxwell (who had worked out the kinetic theory of gases) developed the theory of electromagnetism.

This first identified light as an electromagnetic radiation and predicted the existence of many octaves of such radiation on either side of the visible light range.

In 1888, the German physicist Heinrich Rudolf Hertz (1857–1894) detected lightlike radiation with wavelengths a million times longer than light and with energy levels that were, therefore, only a millionth as high. The new radiation came to be spoken of as radio waves.

Radio waves, *because* of their low energy content, turned out to be easy to produce, and *despite* their low energy content, easy to receive. Radio waves could penetrate all sorts of material objects as light could not. Radio waves could bounce off layers of charged particles in the upper atmosphere as light could not, so that radio waves could, in effect, follow the curve of Earth's surface. Radio waves could easily be produced in coherent fashion, so that a tight beam could go long distances, and could easily be modified to carry messages.

For all these reasons radio waves were clearly ideal for long-range communication, and that, too, without the wires that telegraphs and cables required. The first to make practical use of radio waves in this way was the Italian electrical engineer Guglielmo Marconi (1874–1937). In 1901, he sent a radio-wave signal across the Atlantic Ocean, a feat generally recognized as the invention of radio.

From that day on, with further improvements and refinements, radio became a more and more important means of communication. It was clear to many people that any technological civilization would surely make use of radio communication in preference to anything else.

Therefore, when the planet Mars made a closer than usual approach to Earth in 1924, there was some attempt to listen for radio signals from the presumed civilization that had built its canals. Nothing was detected.

In a way that was not surprising. The layers of charged atoms in the upper atmosphere that reflected Earth-made radio waves and kept them in the neighbourhood of the surface instead of allowing them to pass outwards into space would also serve to reflect space-made radio waves and keep them away from Earth's surface.

In 1931, however, the American radio engineer Karl Guthe Jansky (1905–1950), working for Bell Telephone Laboratories, detected an odd signal when he was trying to determine the source of static that interfered with the developing technique of radio telephony. It turned out that the signal was coming from the sky. That was the first indication that there was a wide band of short-wave radio waves, called microwaves, that could easily penetrate Earth's atmosphere. There were two types of electromagnetic radiations that we could get from the sky: a narrow band of visible light and a broad band of microwaves.

By December 1932, it was demonstrated that Jansky had detected radio waves from the Galactic centre, and that made front-page headlines in the *New York Times*. Some astronomers, such as Jesse Leonard Greenstein (born 1909) and Fred Lawrence Whipple (born 1906), at once appreciated the potentialities of the discovery, but there was little that could be done about it. There were no decent instruments for detecting such radiation. One American radio engineer, Grote Reber (born 1911), did take it seriously, however. He built a device to detect radio waves from the sky (a 'radio telescope') and from his back yard, beginning in 1938, studied as much of the sky as he could reach in order to measure the intensity of radiowave reception from different areas.

During World War II, the development of radar changed everything. Radar made use of microwaves so that microwave technology advanced rapidly, and after the war, radio astronomy quickly became a giant, revolutionizing the science as it had been revolutionized by Galileo's telescope more than three centuries before.

In just a few decades, radio telescopes have been built that can detect microwaves far more delicately than light can be detected. Sources of microwave radiation could be detected at distances too great for us to make out light radiation of anything like equivalent energy. In fact, we can right now detect microwaves from any stars in the Galaxy, even though those microwaves are sent out with no more energy than we ourselves could dispose of.

Then, too, the sources of microwaves can be located with great precision, and the varieties of microwaves can be differentiated with great ease. Every molecule emits or absorbs its own specific wavelength, so that the chemical constitution of interstellar gas clouds can be determined with great precision. Microwaves are not blanked out by background radiation. In most parts of the sky, microwaves are not radiated with the intensity of light, and even where microwaves are plentiful, it would be easy for a civilization to send out a specific wavelength that would be far stronger than the natural background *for that wavelength.*

It amounts to this: If any civilization is trying to send out messages, it would surely come to the conclusion that microwaves are a better, cheaper, and more natural medium for those messages than light – or, in fact, than anything.

We finally have what looks like the answer. To send, or receive, message across the interstellar gulfs, we must make use of microwaves.

But at what energy level, or wavelength, ought we to expect the message to come? Receivers can be tuned to receive some specific wavelength, and if the message is being sent at another wavelength, it will be missed. On the other hand, to try to tune in all possible wavelengths would enormously increase the difficulty and expense of listening. But can we read the extraterrestrial mind and guess the wavelength it would choose to use?

During World War II, the Dutch astronomer Hendrick Chistoffell Van de Hulst (born 1918), unable to make observations under the Nazi occupation, did some pen-and-paper calculations that showed that cold hydrogen atoms would sometimes undergo a change in configuration that would result in the emission of a microwave photon that was 21 centimetres (8.3 inches) in wavelength.

The individual hydrogen atom undergoes the change only

very rarely but, considering all the hydrogen atoms in space, great numbers are undergoing the change at every moment, so that if Van de Hulst's calculations were correct, the microwaves produced by hydrogen atoms should be detectable. In 1951, the American physicist Edward Mills Purcell (born 1912) did detect them.

The hydrogen atom is predominant in the space between the stars, and the 21-centimetre wavelength is therefore a universal radiation that would be received anywhere. Any civilization that had reached our technological level would certainly be radio astronomers, and we can be certain they would have instruments equipped to receive the 21-centimetre wavelength even if they bothered with nothing else. Surely they would transmit messages over a wavelength they could themselves receive and one that they would be certain that all other civilizations would be tuned to.

In 1959, therefore, the American physicist Philip Morrison and the Italian physicist Giuseppe Cocconi (born 1914) suggested that if signals from extraterrestrials were searched for, they should be searched for at 21-centimetre wavelengths.

That is the microwave wavelength, however, in which the background radiation is strongest and potentially the most obscuring – particularly in the region of the Milky Way. There is some feeling, therefore, that we ought to look somewhere else, perhaps at 42 centimetres or 10.5 centimetres, since doubling or halving the obvious choice is the simplest way of using 21-centimetres as the basis for the message without using that wavelength itself.

Another suggestion is to make use of hydroxyl, the 2-atom combination of hydrogen and oxygen, which, next to hydrogen itself, is the most widespread emitter of microwaves in interstellar space. Its microwave emission has a wavelength of 17 centimetres (6.7 inches).

Since hydrogen and hydroxyl together make water, the stretch of microwaves from 17 to 21 centimetres in wavelength is sometimes called the waterhole. The name is particularly apt, because the hope is that different civilizations will send and receive messages in this region as different species of animals come to drink at literal waterholes on Earth.

In 1960, the first real attempt was made to listen to the 21-centimetre wavelength in the sky in the hope of detecting messages from extraterrestrial civilization. It was carried through in the United States under the direction of Frank Drake, who called it Project Ozma. Ozma was Princess of Oz, the distant land in the sky of the well-known children's adventure series. After all, the astronomers were trying to gain evidence of occupied lands even farther in the sky than Oz is.

The listening began at 4 a.m. on 8 April 1960, with absolutely no publicity, since the astronomers feared ridicule. It continued for a total of 150 hours through July, and the project then came to an end. The listeners were on the alert for anything with a very narrow range of wavelengths that seemed to flicker in a way that was neither quite regular nor quite random. They detected nothing of the sort.

Since Project Ozma, there have been six or eight other such programmes, all at a level even more modest than the first, in the United States, in Canada, and in the Soviet Union. There have been no positive results, but the fact is that the search has been very brief and superficial so far.

Astronomers remain alive to the possibility of accidental discoveries, of course. When, in 1967, pulsars (very tiny, very dense, very rapidly rotating stars that were remnants of collapse following supernova explosions) were discovered, for just a short while the surprising detection of pulses of microwaves gave the astronomers concerned an eerie feeling that messages of intelligent origin were being received. They referred to it as the LGM ('little green men') phenomenon. The pulses quickly proved far too regular to be carrying a message, however, and less dramatic explanations were found.

If the search for messages from extraterrestrial civilizations is to be carried through with some reasonable hope of success, however, far more time must be spent than was the case in

Project Ozma; for more stars must be studied, far more elaborate equipment must be used. In short, a very expensive project must be set up.

Where?

In 1971, a NASA group under Bernard Oliver suggested what has come to be called Project Cyclops.

This would be a large array of radio telescopes,* each 100 metres (109 yards) in diameter, and each adjusted for reception of microwaves in the waterhole region.

The array would consist of 1,026 such radio telescopes in rank and file, all of them steered in unison by a computerized electronic system. The entire array working together would be equivalent to a single radio telescope some 10 kilometres (6.2 miles) across.

The array would be capable of detecting something as weak as Earth's inadvertent leakage of microwaves even from a distance of 100 light-years, while the deliberately emitted message beacon of another civilization could be detected at a distance of at least 1,000 light-years.

Earth's surface may not be the best place for it. If it could be built in space, or, better yet, on the far side of the Moon, it would be insulated from most or all of the background of Earth's own microwave noise.

Project Cyclops would not be easy to construct and certainly not cheap. Estimates are that the construction and maintenance of the array and the search itself would cost anywhere from ten to fifty billion dollars, even allowing for the fact that eventually the listening will be completely computerized and will not take much in the way of people-hours.

Anything that could be done to make the search simpler and quicker would be helpful, therefore. There might be places in

* Each radio telescope would seem like a round eye, metaphorically speaking, gazing at the sky. The word *cyclops* is Greek for *round eye*.

the sky, for instance, where it would pay us to search first because they are more likely sources of messages than other places are.

Where might these places be?

First, the best place to search is in the neighbourhood of some star where a planetary civilization with copious energy at its disposal might exist. (There might be, to be sure, signals being sent out by free-worlds or automatic probes that are closer to us than any star, but we have no way of knowing where such objects are and therefore no particular target to aim at.)

Second, the objective should be a nearby star rather than a distant star, since, all things being equal, the microwave beam will be more intense and easier to detect the closer the planetary system from which it starts.

Third, the objective should be a Sunlike star, since it is there we expect habitable planets might exist.

Fourth, the first objectives should be single stars, since, even though it seems that binary stars may still have habitable planets circling them, the chances are perhaps greater in the case of single stars.

As it happens, there are just seven Sunlike single stars within two dozen light-years of us, and they are:

Star	Distance (light-years)	Mass (Sun = 1)
Epsilon Eridani	10.8	0.80
Tau Ceti	12.2	0.82
Sigma Draconis	18.2	0.82
Delta Pavonis	19.2	0.98
82 Eridani	20.9	0.91
Beta Hydri	21.3	1.23
Zeta Tucanae	23.3	0.90

None of these stars has a familiar name, for those that do are generally the brightest, which are too large and short lived to be suitable for civilizations.

Stars that are visible to the unaided eye, even if they are not outstandingly bright, are generally named for the constellation in which they are found. Sometimes they are listed in order of brightness, or position, by the use of Greek letters (alpha, beta, gamma, delta, epsilon, zeta, and so on) or by Arabic numerals.

The stars in the table above are from the constellations Eridanus (the River), Cetus (the Whale), Draco (the Dragon), Pavo (the Peacock), Hydrus (the Water Snake), and Tucana (the Toucan).

Of the seven stars listed in the table, three – Delta Pavonis, Beta Hydri, and Zeta Tucanae – are located so far south in the sky as to be invisible from the northern climes where astronomy is most advanced and where complex equipment exists in the greatest profusion. As for 82 Eridani, that is not too far south to be visible, but it is apt to be too near the horizon for complete comfort.

The three very best targets, then, are Epsilon Eridani, Tau Ceti, and Sigma Draconis. Project Ozma, at the suggestion of the Russian-American astronomer Otto Struve, concentrated on Epsilon Eridani and Tau Ceti.

Although these seven stars, and particularly the three northern stars, are the obvious targets for the first phase of the search, we should not quit if the results are negative. If there are seven prime targets within twenty-three light-years, there would be about 500,000 altogether within the 1,000-light-year reach of the Project Cyclops array.

Ideally, we should listen to all of them. In fact, before we really give up hope, we should scan the entire sky, just in case civilizations are present in the neighbourhood of surprising stars – or just in case we get signals from probes or free-worlds that are fairly close to us without our being aware of it.

We should even search wavelength ranges outside the water-hole, just in case.

Why?

Yet one must ask: Why ought humanity to engage in the task of monitoring space for signals from extraterrestrial civilizations? Why should we spend tens of billions of dollars when the chances are that we may find nothing at all?

After all, what if, despite all my reasoning in this book, there are no extraterrestrial civilizations?

– Or if there are, that there are none so close to us that we can detect their signals?

– Or if there are, that they are not signalling?

– Or if they are, that they are doing so in a way that will elude us altogether?

– Or if it doesn't, that the signals we receive will be uninterpretable?

Any of these things is possible, so let us assume the worst and suppose that despite all our efforts, we end up with no recognizable signals at all from anywhere.

In that case, will we really have wasted much money?

Perhaps not. Suppose that the labour of building Project Cyclops and the task of searching the sky takes twenty years altogether and costs 100 billion dollars. That is five billion dollars a year in a world in which the various nations spend a total of 400 billion dollars a year on armaments.

And whereas the money spent on armaments only stimulates hatred and fear and increases steadily the chance that the nations of the Earth will wipe out each other and, perhaps, all humanity, the search for extraterrestrial intelligence is something that would surely have a uniting effect on us all. The mere thought of other civilizations advanced beyond our own, of a Galaxy full of such civilizations, can't help but emphasize the pettiness of our own quarrels and shame us into more serious attempts at cooperation. And if the failure of the search should cause us to suspect that we are, after all, the only civilization in the Galaxy, might that not increase the sense of the preciousness of our world and ourselves and make us more reluctant to risk it all in childish quarrels?

But will the money be wasted at all if we end up with nothing?

In the first place, the very attempt to construct the equipment for Project Cyclops will succeed in teaching us a great deal about radiotelescopy and will undoubtedly advance the state of the art greatly even before so much as a single observation of the heavens is made.

Secondly, it is impossible to search the heavens with new expertise, new delicacy, new persistence, new power, and fail to discover a great many new things about the Universe that have nothing to do with advanced civilizations. Even if we fail to detect signals, we will not return from the task empty-handed.

We can't say what discoveries we will make, or in what direction they will enlighten us, or just how they may prove useful to us, but humanity has (at its best moments) always valued knowledge for its own sake. The ability to do that is one of the ways in which a more intelligent species would be differentiated from a less intelligent one; and an advancing culture is differentiated from a decaying one.

Nor need we fear that in the end knowledge will have to be valued for its own sake only. Knowledge, wisely used, has always been helpful to humanity in the past; and there is every hope it will continue to be helpful in the future.

But suppose we do find a signal of some sort and decide that it must be of intelligent origin. Will that be of great value to us?

It may be that it won't be a beacon at all; that no one is trying to attract our attention or to tell us anything. It may be the inadvertent overflow of technology, just a jumble of everyday activity, like the ball of microwaves that is now steadily expanding from the Earth in every direction.

That in itself – the mere recognition of a signal as representing the existence of a far-off civilization, even one from which we can extract no information at all – is quite enough, in some ways.

Think of the psychological significance right there. It means that somewhere else a civilization exists, which, judging from the mere strength of its signals, might just be advanced beyond our own.* That alone gives us the heartening news that at least one group of intelligent beings has reached our level of technology and has succeeded in not destroying itself, but has instead survived and advanced onward to greater heights. And if they have done so, may we not do so as well?

If this thought helps keep us from despair during humanity's mountainous tasks of solving the problems that lie immediately ahead of us, that alone may help move us towards the solution. It might even, perhaps, provide the crucial feather's weight that may swing the balance towards survival and away from destruction.

Nor can it be possible that we will get no information other than the mere existence of the signal. Even if there is no intelligent message in the signal, or none that we can interpret, the characteristics of the signal could tell us the rate at which the signal-sending planet revolves about its star and rotates about its axis, together with perhaps other physical characteristics that could be of great interest and use to astronomers.

And suppose we recognize that there *is* useful material in the message, yet remain at a total loss to determine what that useful material might mean.

Is the message then useless? Of course not. In the first place, it presents us with an interesting challenge, a fascinating game in itself. Without coming to any conclusion as to specific items of information, we might reach certain generalizations concerning alien psychologies – and that, too, is knowledge.

Besides, even the tiniest breaks in the code could be of interest. Suppose, for the sake of argument, that from the message we get the hint of a relationship that, if true, might give

* On the other hand, if we detect nothing, that is not definitive proof that there is nothing there. We may be looking in the wrong place, or in the wrong fashion, or with the wrong technique, or all three.

us a new insight into some aspect of physics – it might even seem a trivial insight. Yet scientific advances do not exist in a vaccum. That one insight might stimulate other thoughts and, in the end, greatly accelerate the natural process by which our scientific knowledge advances.

And if we do come to some detailed understanding of the message, we might learn enough to be able to deduce whether the civilization sending it is peaceful or not.

If it is dangerous and warlike (a very slim chance, in my opinion), then the knowledge we will have gained will encourage us to keep quiet, make no reply, do our best to shield as far as possible any leakage into outer space of anything that will give a hint of our presence. Perhaps the knowledge we gain will give us some insight into how best to defend ourselves if the worst comes to the worst.

If, on the other hand, we decide that the messages are coming from a peaceful and benign civilization, or from one that cannot reach us whatever its attitude, then we might decide to answer, using the code we have learned.

To be sure, the civilization may be so far away from us that, thanks to the speed-of-light limit, we cannot expect an answer for, say, a century. There is, however, no great problem in waiting. We can go about our own business while we wait, so we lose nothing.

The advanced civilization at the other end, on receiving our answer and knowing that someone is listening, may perhaps at once begin to transmit in earnest. Though we wait a century for it, we would find ourselves thereafter getting a cram-course in all aspects of the alien civilization.

There is no way we can predict how useful such information will prove to be, but surely it cannot be useless.

In fact, if we move to the romantic extreme of supposing that the speed-of-light limit can be beaten and that there is a peaceful and benign Federation of Galactic Civilizations, our successful interpretation of the message and our courageous answer may amount to our ticket of entrance.

Who knows?

*

Even disregarding the vast curiosity that has always driven humanity, and the intense interest we all must have in so overwhelming a question as to whether or not there are other civilizations in the Universe in addition to our own, it does seem to me that no matter what we do in attempting to answer that question, we will succed in profiting and in helping ourselves.

Therefore, for the sake of all of us, let's abandon our useless, endless, suicidal bickering and unite behind the real task that awaits us – to survive – to learn – to expand – to enter into a new level of knowledge.

Let us strive to inherit the Universe that is waiting for us; doing so alone, if we must, or in company with others – if they are there.

Index

Index

Guy Lyon Playfair and Scott Hill
The Cycles of Heaven £1.20

Why do so many suicides occur in April and May? Can the appearance of sunspots be linked with how people vote? What *really* happened to the Soyuz XI astronauts? What secrets are to be found in an electromagnetic wave? Are world-shaking upheavals in store for us in the early 1980s? This book considers the serious possibility that the fortunes of man oscillate in cycles – driven by our all-powerful sun.

'A serious work and a significant one' BRIAN INGLIS, DAILY MAIL

Francis Hitching
The World Atlas of Mysteries £5.50

From the origins of the universe and terrestrial life, through the unique development of man, to the secrets of ancient civilizations and bizarre phenomena in the sky and beyond – the enormous scope of this encyclopedia, its exhaustive research and copious illustrations (maps, photographs, diagrams) make it a unique and fascinating book. Francis Hitching, author of *Earth Magic*, is one of the world's leading authorities on the inexplicable and the unexplained.

'A book of absorbing interest to anyone who believes that there are more things in heaven and earth than science will recognize' DR KIT PEDLER (creator of DOOMWATCH), EVENING NEWS

Brandt Commission
North-South £1.95
a programme for survival

The result of a unique and independent investigation by
world statesmen into the urgent problems of inequality
faced by a world economic system which favours the prosperous
countries of the 'North' at the cost of the poorer 'South'. The
report of the Independent Commission on International Issues
under the chairmanship of Willy Brandt argues that major initiatives
are needed if mankind is going to survive: long-term reforms by the
year 2000, priority programmes for the 1980s, and emergency
action to avert an imminent crisis. It calls on all countries to make
an imaginative response to problems which affect us all.

Alvin Toffler
Future Shock £1.95

Future shock is the disease of change. Its symptoms are
already here ... *Future Shock* tells what happens to people
overwhelmed by too rapid change ... And looks at the human side
of tomorrow. Brilliantly disturbing, the book analyses the new and
dangerous society now emerging, and shows how to come to
terms with the future.

'An important book reaching some startling conclusions' BBC

'If this book is neglected we shall all be very foolish' C. P. SNOW

Magnus Pyke
Butter Side Up! 75p

Could we soon be eating dormice for dinner? Does bread really fall
more often butter side up ... or butter side down? Britain's most
popular scientist asks some not-so-silly questions and comes up with
some very sensible answers.

' ... achieves the small miracle of evoking in print Dr Pyke's
now familiar torrent of talk' SUNDAY TELEGRAPH

Bestselling Fiction and Non-Fiction

☐ Fletcher's Book of Rhyming Slang	Ronnie Barker	80p
☐ Pregnancy	Gordon Bourne	£2.50p
☐ A Sense of Freedom	Jimmy Boyle	£1.25p
☐ The Thirty-nine Steps	John Buchan	80p
☐ Out of Practice	Rob Buckman	95p
☐ The Flowers of the Forest	Elizabeth Byrd	£1.50p
☐ The 35mm Photographer's Handbook	Julian Calder and John Garrett	£6.95p
☐ Women Have Hearts	Barbara Cartland	70p
☐ The Sittaford Mystery	Agatha Christie	85p
☐ Lovers and Gamblers	Jackie Collins	£1.50p
☐ Sphinx	Robin Cook	£1.25p
☐ The Life	Jeanne Cordelier	£1.50p
☐ Soft Furnishings	Designers' Guild	£5.95p
☐ Rebecca	Daphne du Maurier	£1.25p
☐ Peter Finch	Trader Faulkner	£1.50p
☐ The Complete Calorie Counter	Eileen Fowler	50p
☐ The Diary of Anne Frank	Anne Frank	85p
☐ Flashman	George MacDonald Fraser	£1.25p
☐ Wild Times	Brian Garfield	£1.95p
☐ Linda Goodman's Sun Signs	Linda Goodman	£1.95p
☐ The 37th Pan Book of Crosswords	Mike Grimshaw	70p
☐ The Moneychangers	Arthur Hailey	£1.50p
☐ The Maltese Falcon	Dashiell Hammett	95p
☐ Vets Might Fly	James Herriot	95p
☐ Simon the Coldheart	Georgette Heyer	95p
☐ The Eagle Has Landed	Jack Higgins	£1.25p
☐ The Seventh Enemy	Ronald Higgins	£1.25p
☐ To Kill a Mockingbird	Harper Lee	£1.25p
☐ Midnight Plus One	Gavin Lyall	£1.25p
☐ Chemical Victims	Richard Mackarness	95p
☐ Lady, Lady, I Did It!	Ed McBain	90p
☐ Symptoms	edited by Sigmund Stephen Miller	£2.50p
☐ Gone with the Wind	Margaret Mitchell	£2.95p
☐ Robert Morley's Book of Bricks	Robert Morley	£1.25p

☐	**Modesty Blaise**	Peter O'Donnell	95p
☐	**Falconhurst Fancy**	Kyle Onstott	£1.50p
☐	**The Pan Book of Card Games**	Hubert Phillips	£1.25p
☐	**The New Small Garden**	C. E. Lucas Phillips	£2.50p
☐	**Fools Die**	Mario Puzo	£1.50p
☐	**Everything Your Doctor Would Tell You If He Had the Time**	Claire Rayner	£4.95p
☐	**Polonaise**	Piers Paul Read	95p
☐	**The 65th Tape**	Frank Ross	£1.25p
☐	**Nightwork**	Irwin Shaw	£1.25p
☐	**Bloodline**	Sidney Sheldon	95p
☐	**A Town Like Alice**	Nevil Shute	£1.25p
☐	**Lifeboat VC**	Ian Skidmore	£1.00p
☐	**Just Off the Motorway**	John Slater	£1.95p
☐	**Wild Justice**	Wilbur Smith	£1.50p
☐	**The Spoiled Earth**	Jessica Stirling	£1.75p
☐	**That Old Gang of Mine**	Leslie Thomas	£1.25p
☐	**Caldo Largo**	Earl Thompson	£1.50p
☐	**Future Shock**	Alvin Toffler	£1.95p
☐	**The Visual Dictionary of Sex**	Eric J. Trimmer	£5.95p
☐	**The Flier's Handbook**		£4.95p

All these books are available at your local bookshop or newsagent, or
can be ordered direct from the publisher. Indicate the number of copies
required and fill in the form below

Name_____
(block letters please)

Address_____

_____

Send to Pan Books (CS Department), Cavaye Place, London SW10 9PG
Please enclose remittance to the value of the cover price plus:

25p for the first book plus 10p per copy for each additional book ordered
to a maximum charge of £1.05 to cover postage and packing
Applicable only in the UK

While every effort is made to keep prices low, it is sometimes
necessary to increase prices at short notice. Pan Books reserve
the right to show on covers and charge new retail prices which
may differ from those advertised in the text or elsewhere